Particle Technology

Powder Technology Series

EDITOR BRIAN SCARLETT

Delft University of Technology
The Netherlands

Many materials exist in the form of a disperse system, for example powders, pastes, slurries, emulsions and aerosols. The study of such systems necessarily arises in many technologies but may alternatively be regarded as a separate subject which is concerned with the manufacture, characterization and manipulation of such systems. Chapman and Hall were one of the first publishers to recognize the basic importance of the subject, going on to instigate this series of books. The series does not aspire to define and confine the subject without duplication, but rather to provide a good home for any book which has a contribution to make to the record of both the theory and the application of the subject. We hope that all engineers and scientists who concern themselves with disperse systems will use these books and that those who become expert will contribute further to the series.

Particle Size Measurement
T. Allen
4th edn, hardback (0 412 35070 X), about 736 pages

Powder Porosity and Surface Area
S. Lowell and Joan E. Shields
2nd edn, hardback (0 412 25240 6), 248 pages

Pneumatic Conveying of Solids
R. D. Marcus, L. S. Leung, G. E. Klinzing and F. Rizk
Hardback (0 412 21490 3), 596 pages

Particle Technology
Hans Rumpf
Translated by F. A. Bull
Hardback (0 412 35230 3), about 216 pages

Particle Technology

HANS RUMPF
*Late Professor of Mechanical Process Technology
at the University of Karlsruhe*

Translated by
F. A. BULL M.Sc. (Melb), Dr.rer.nat. (Karlsruhe)
*Formerly Senior Lecturer, Department of Chemical Engineering,
University of Melbourne*

CHAPMAN AND HALL
LONDON•NEW YORK•TOKYO•MELBOURNE•MADRAS

UK   Chapman and Hall, 11 New Fetter Lane, London EC4P 4EE
USA   Chapman and Hall, 29 West 35th Street, New York NY10001
JAPAN   Chapman and Hall Japan, Thomson Publishing Japan,
                Hirakawacho Nemoto Building, 7F, 1-7-11 Hirakawa-cho,
                Chiyoda-ku, Tokyo 102
AUSTRALIA   Chapman and Hall Australia, Thomas Nelson Australia, 480 La
                Trobe Street, PO Box 4725, Melbourne 3000
INDIA   Chapman and Hall India, R. Sheshadri, 32 Second Main Road,
                CIT East, Madras 600 035

Original German language edition (Mechanische Verfahrenstechnik)

© 1975 Carl Hanser Verlag, Munich/FRG
Softcover reprint of the hardcover 1st edition 1975
English edition 1990
© Carl Hanser Verlag, Munich/FRG
Authorized translation from the German language edition
published by Carl Hanser Verlag, Munich/FRG.

Typeset in 10/12 Times by
KEYTEC, Bridport, Dorset

ISBN-13: 978-94-011-7946-1

British Library Cataloguing in Publication Data

Rumpf, Hans
Particle technology.
1. Particle technology
I. Title   II. Series
620'.43

ISBN-13: 978-94-011-7946-1

Library of Congress Cataloging-in-Publication Data

Rump, Hans,1911-
    [Mechanische Verfahrenstechnik. English]
    Particle technology/Hans Rumpf; translated by F. A. Bull.
       p.   cm.–(powder technology series)
    Translation of: Mechanische Verfahrenstechnik.
    Includes bibliographical references.
    ISBN-13: 978-94-011-7946-1   e-ISBN-13: 978-94-011-7944-7
    DOI: 10.1007/978-94-011-7944-7
    1. Particles. I. Title. II. Series.
    TP156.P3R8613   1990
       660.2—dc20   89-48692
                                                              CIP

Contents

Acknowledgements

This monograph was written with the assistance of Messrs W. Gleißle, H. P. Kurz, S. Moser, M. H. Pahl, K. H. Sartor, G. Schädel, K. Sommer; Drs B. Koglin, J. Raasch, H. Reichert, H. Schubert; and Prof. K. Schönert, all of the University of Karlsruhe.

Translator's preface

The inspiration for translating this classic text came during a sabbatical year spent at the University of Karlsruhe in 1974.

Under the leadership of the late Professor Hans Rumpf, the Institut für Mechanische Verfahrenstechnik, Karlsruhe, from the early 1960s onwards, by extensive research and advanced teaching had promoted the discipline of mechanical process technology, a branch of process engineering which had been rather neglected, especially in many chemical engineering departments of universities in the English-speaking world.

There is a need for texts of this kind, particularly for the more specialized teaching that has to be done during the later stages of engineering courses. This work, which is really a monograph, serves as a concise and compact introduction, albeit at an advanced level, to all those functions of process engineering that have to do with the handling and treatment of particulate matter and bulk solids. Much of this information has previously been scattered around journals and other books and not brought together in one work. Furthermore, Rumpf has emphasized the physical and theoretical foundations of the subject and avoided a treatment that is simply empirical.

I am indebted to two of his former colleagues, Professor Klaus Schönert and Professor Friedrich Löffler, for supporting me in this translating venture, and especially to Dr John C. Williams, former Dean of Engineering at the University of Bradford, who has edited the English text and at whose university I began the translation some years ago. I am also much obliged to Mr John C. Barton for a very thorough reading of the proofs.

F. A. Bull
Melbourne, March 1989

Preface

Mechanical processes for transforming materials, such as milling, mixing, agglomerating, sedimenting, the winnowing of grain etc., are among the oldest useful arts practised by mankind. At the beginning of the modern industrial era extensive knowledge and experience of many industrial applications, e.g. in mining, in the manufacture of ceramics, chemicals and textiles, in the production of paper and in food technology, had already been accumulated. This practical knowledge was substantially broadened in the course of technical developments, especially as many completely new processes came into being and the traditional methods were adjusted to the advancing state of technology. Science concerned itself with the systematic understanding and design of processes and equipment. At the same time the concept of **unit operations** was pursued whereby the whole field is divided into a larger number of basic processes that are defined according to their purposes. The principal groups into which these processes fall are those of separating and mixing without any intentional changes in particle size, and those where the particle size is altered, as in comminution and grain enlargement. In addition, there are the procedures of conveying, dumping and feeding. This classification is also followed in the survey of these processes given in Chapter 4 of this monograph.

Empiricism still plays a large part in the practical application of these unit operations, i.e. basic processes. Consequently the scientific expositions are in many cases merely descriptive. However, the value of the information obtainable from comprehensive descriptions of processes is limited. For the solution of particular problems more exact information is necessary than can be provided by a detailed descriptive presentation of the subject which, in its technical features, is extraordinarily complex.

The essential unity of the subject would also not become apparent in a predominantly descriptive account of the basic processes. This unity rests on common scientific foundations. Up to now these have been dealt with only in individual publications and, to some extent, in books which have concerned themselves with the whole of process technology or branches thereof.

However, the foundations of mechanical process technology have not yet been collectively promulgated. They have been elaborated during the period 1960–1974 in a course of lectures entitled 'Fundamentals of Mechanical Process Technology', on which the teaching of mechanical process technology at the University of Karlsruhe has been based. In these lectures the theoretical and physical foundations of the subject have been dealt with according to a unified point of view, i.e. not bound up with technically and practically oriented unit operations. These unit operations are the concern of other, subsequent, lecture courses which build upon the fundamental lectures.

The chapter 'Mechanical Process Technology' from *Chemical Technology* by Winnacker and Küchler [29, Vol. 7], which appears here as a monograph, emphasizes the scientific fundamentals and thereby follows the above-mentioned lecture course.

Of course, there is not enough room to permit a more extensive explanatory presentation; this must be reserved for a textbook. A solution has been sought in the form of a compromise that ought to meet the requirements of a handbook. In Chapters 2 and 3, whose purpose is to present the fundamentals, the main questions are addressed and their theory is presented in sufficient detail to enable the methodology to be grasped and understanding of the subject to be pursued. In many cases the derivation of formulae has been dispensed with. The reader who is more interested in a general survey of the subject and less interested in its actual mathematical presentation can ignore the formulae. Even then he will be able to obtain an insight into the questions arising in mechanical process technology and into its scientific development.

I wish to thank numerous colleagues in my department for discussions and practical help. They are mentioned in the particular chapters. Messrs W. Gleißle and G. Schädel were responsible for the formal completion of the manuscript and the preparation of the figures, and rendered valuable help.

1
Introduction

1.1 THE DEFINITION OF MECHANICAL PROCESS TECHNOLOGY

Mechanical process technology deals with the transformation of material systems by predominantly mechanical operations. It embraces the whole field of particle technology, by which we mean the transformation and transport of particulate systems susceptible to change by mechanical means. The size of the particles in such systems extends from about 10 m down to the colloidal range, although with decreasing particle size the mechanical operations possible become less effective in comparison with thermal, electrical and chemical influences; in other words, this technology is concerned with the whole range of particulate, i.e. disperse, systems from the coarse down to the colloidal.

The elements of the disperse systems of mechanical process technology are generally particles which are distinguishable from one another. In addition to such systems, mechanical process technology is also concerned with those systems in which no discernible particles occur but where continuous phases mutually permeate one another as is the case, for example, in coloured plasticine. The solid–pore system of a sponge is also of this type. The flow through such a system of pores is still a matter for mechanical process technology because it is no different in principle from the flow through a porous bed of discrete particles.

Mechanical process technology is also concerned with mixing, whether this has to do with phases consisting of discrete particles, continuous fluid phases or mutually soluble phases, i.e. mixing on the molecular scale.

Single-phase flow, as such, is a matter for fluid mechanics; mechanical process technology uses this as a basis. To deal scientifically with mechanical process technology, a knowledge of the fundamental principles governing the flow of Newtonian fluids must be assumed as well as a knowledge of the mechanics of solids. Rheology, which is also a science in its own right, provides in its theoretical structure the overall basis for the mechanics of elastic and flowable materials. This does not prevent mechanical process technology from concerning itself with certain rheological topics such as the flow of non-Newtonian fluids, the rheology of suspensions or the flow of bulk solids.

The important changes of state and transport phenomena of the various branches of process engineering are set out schematically in Fig. 1.1. The important transformations in chemical process technology are those brought about by chemical reactions; in thermal process technology they are changes of phase and heat and mass transfer between thermodynamically defined phases. In both cases the states of equilibrium towards which the systems tend to move are determined by the laws of thermodynamics. These states of equilibrium depend thermodynamically upon the intensive variables, pressure and temperature, and the composition of the system, and not directly upon any motion imparted to it. In mechanical process technology the changes in the state of aggregation of systems, ranging from coarse to colloidal, consist of alterations in the degree of dispersity and the extent of mixing of such disperse systems with continuous fluid phases. The alteration in the state of dispersity can take the form of the nebulization of liquids or the sifting of solid particles. It is evident that the state reached

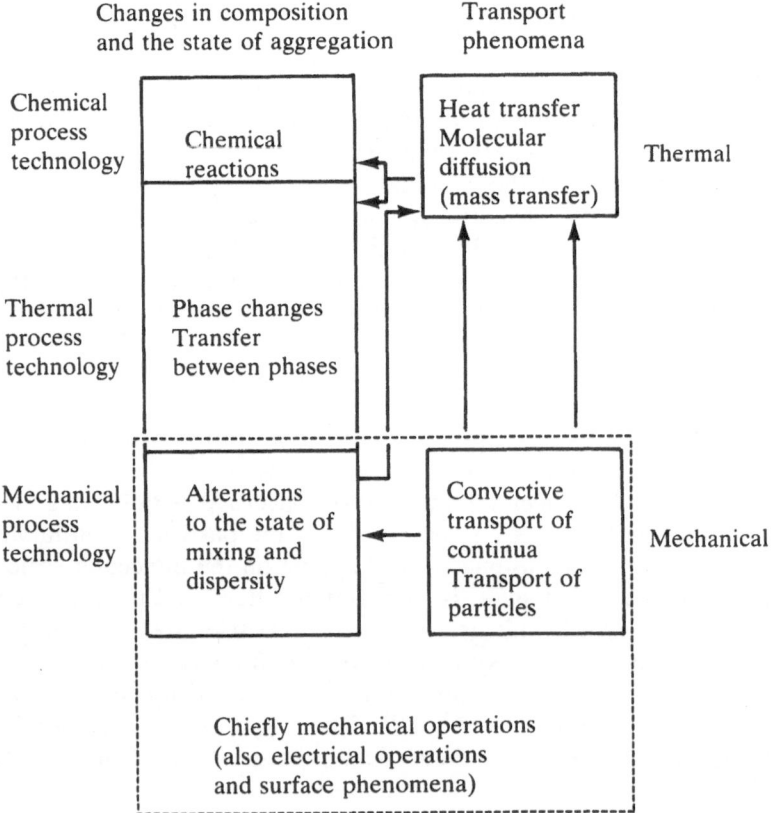

Figure 1.1 Changes of state and transport phenomena in process engineering.

by the system depends upon the forces exerted on its elements and the motion imparted to these. Thus the state of the system does not change in a direction determined by its intensive properties, as these are independent of its motion.

1.2 THE PURPOSE OF MECHANICAL PROCESS TECHNOLOGY

Mechanical processes first enable chemical and thermal processes to be carried out effectively, and they frequently precede thermal and chemical processes or are often directly associated with them. Indeed, the beneficial effects of mechanical treatment for chemical and thermal processes can first ensue when the product is used by the consumer. For example, cement clinker is ground at the cement works to a particular particle size distribution so that the concrete mixture used on a building site will attain the necessary strength within the setting time allowed. In other cases mechanical processes are important in their own right for obtaining products with the required properties. Sugar as lump sugar, ordinary sugar crystals or castor sugar should dissolve sufficiently quickly, i.e. possess thermal reactivity. At the same time, however, properties suiting particular applications are demanded which are associated with sugar cubes or coarse crystals or castor sugar or with powdered sugar in the form of coarse granules.

Some properties of disperse systems which depend upon particle size are listed in Table 1.1. For example, chemical and mineralogical homogeneity is of supreme importance in ore dressing or flour milling. Chemical and thermal reactivity have already been referred to. When using abrasives, a certain grinding performance must be achieved; however, in the case of optical glasses, for example, a surface whose structure is uniform and free from scratches is required. White pigments should reflect incident light as diffusely and completely as possible and must have no coarse grains which impair the gloss of the surface finish. With coloured pigments there must be a definite ratio of the amount of light selectively absorbed to the amount scattered. Insecticides should, like pollen, be as dusty as possible but not so fine that they agglomerate. The sensation of taste is strongly affected by the fineness of particles; for example, sugar particles larger than $25\,\mu\text{m}$ make chocolate seem gritty and sugar which is too fine makes the chocolate unpleasantly sweet and viscid, whereas cocoa particles which are ground as fine as possible give their full aroma to the cocoa butter. Particle size also influences the flowability of the chocolate paste. On economic grounds alone various properties always compete with one another. The loose demand, made all too often, that a material should be 'as fine-grained as possible' is inadmissible. With increasing fineness the

Table 1.1 Properties of disperse systems which depend upon particle size

Properties of single particles

1   Homogeneity (increasing):
     chemical homogeneity, e.g. in minerals and ores; mineralogical homogeneity, when various crystal modifications exist; physical homogeneity, i.e. homogeneity with regard to certain physical properties. Distinction between properties unaffected by structure, e.g. elasticity, specific heat capacity, optical absorption and properties which are affected, e.g. plasticity, strength
2   Elastic–plastic behaviour (usually increased ductility)
3   Probability of breakage (decreasing), strength (increasing)
4   Wear behaviour, suitability for mechanical surface treatment
5   Properties resulting from competition between volume and surface forces, e.g. adhesion, agglomeration, suspendability, mobility in an electric field (increasing)

Properties of particle assemblies

1   Bulk density (decreasing)
2   Rheological behaviour:
     elasticity, yield point, internal friction, viscosity of a suspension (mostly increasing)
3   Trickling capability (decreasing), flow properties
4   Miscibility (first increasing, then decreasing)
5   Separability (decreasing)
6   Wetting (decreasing)
7   Capillary pressure in solid–liquid systems (increasing)
8   Strength of agglomerates and briquettes (increasing)
9   Fluid and particle mechanics:
     permeability, fluidizability, settling velocity of clusters
10   Thermal properties:
     heat and mass transfer
11   Ignitability, explosive behaviour (increasing)
12   Properties as regards taste
13   Optical properties (extinction, diffuse reflection)

The direction in which the properties tend to change with decreasing particle size is given in parentheses.

cost of milling becomes progressively greater. In addition, other properties change in such a way that, in practice, they have very complicated and expensive, if not prohibitive, consequences. For example, the tendency to agglomerate and stick to walls increases, the flow properties become extremely poor, the permeability decreases and separation from air and water becomes much more difficult.

In formulating what mechanical process technology is about, it is thus a matter of knowing in each case how the product properties of interest depend upon those properties of particulate systems which can be changed by mechanical means, such as the particle size distribution, the concentration, the state of mixing, the state of agglomeration etc. Opposing the role of these properties is that of the factors representing the costs of producing the desired degree of dispersity. Furthermore, if indices can be specified which quantify the value of the product's properties, then an optimal solution to a given task in mechanical process technology can be found with regard to the expense entailed. Usually the basis for such an optimization is not available. In particular, much effort is required to measure the effects of the properties because, for this purpose, it is necessary to produce size fractions of the products involved and various mixtures of such fractions and to measure the properties of these products. That is often worthwhile only for mass production or with very expensive materials.

1.3 MAIN TOPICS AND PROCESSES

The **changes of state** which are brought about mechanically in disperse systems can be divided into two main groups: changes in which the particles of the system alter in size, shape and surface properties; changes in which the particles remain unaltered or do not participate.

The physical changes in the first group are the province of solid state physics, the mechanics of solids and the physics and chemistry of surfaces; those of the second group are the province of fluid mechanics and the kinematics of rigid bodies. The rigid and yield behaviour of rheological systems which can possess both solid and fluid properties, e.g. high polymers or cohesive bulk solids, falls in a transitional region. Following the conventional division of process engineering into unit operations, the first group comes under comminution and grain enlargement and the second group under separation and mixing. Also involved is the handling of continuous media, e.g. the stirring of liquids or the kneading of highly viscous masses.

If we look at the individual operations of a technical process, we notice that all the above-mentioned physical changes of state can occur in each operation. For example, agglomeration occurs during comminution and, conversely, comminution often takes place during agglomeration, e.g. in briquetting. Dynamic processes obviously play an important role in both comminution and agglomeration. The operations of separating and mixing are frequently and often quite decisively influenced by agglomerating and comminuting phenomena.

Included amongst the **transport phenomena** of mechanical process technology, considered more broadly, are storing, feeding and discharging, dosing, and packaging and conveying. All operations present their particular difficulties. The lengthy storage of powdery materials, or of granular materials such as grain, in large silos and storage sheds demands constant or periodic handling of the product because otherwise the condition of the material will change through agglomeration, warming and chemical reactions.

The flow properties of a bulk material must be considered when feeding it into and removing it from a device or storage bin. Special problems occur with the transport of granular material against a larger pressure difference or out of a monte-jus when the material is not sufficiently self-sealing. Likewise the packaging and feeding of granular materials is quite decisively influenced by the flow properties and, in addition to these, by the packing properties such as the bulk volume and the homogeneity and stability of the poured material.

Particle size measurement is a technique of mechanical process technology. It is used to measure the state of the system with regard to its disperity, i.e. to determine particle size distributions, specific surface areas, dust contents, particle velocities, specific material properties of particles and the like. The use of particle size measurement, along with the measurement of other properties, must not be avoided since otherwise essential information about the system under consideration would be unavailable. Particle size distributions, dust contents etc. are generally much more difficult to determine than pressures or temperatures. That is why the measurement of particle size demands considerable effort and great care.

Only a few processes in mechanical process technology work with such low concentrations of particles that the mutual interactions between them can be neglected. In practice such processes are the separation of particles from aerosols and the removal of solids in the treatment of feed water for boilers and water for drinking As a general rule the mutual interactions of particles with one another, with a fluid and with the walls of the apparatus play an important role. The laws governing these mutual interactions are complicated and for the most part not yet properly understood. Therefore many practical applications cannot be designed exactly in advance on the basis of an understanding of the actual physical phenomena. This also explains why many processes have been developed empirically, why the use of more demanding scientific methods is avoided in practice and why such methods are not used to an extent corresponding to the technical and economic importance of the mechanical process. This attitude is thought to be justified on the grounds that scientific research has preferred to investigate the theoretical aspects of simpler problems, e.g. the motion of isolated individual particles in a fluid, which are of only limited interest for direct practical application.

Scientific research, for its part, must pursue two paths: first the systematic phenomenological study of practical processes and the collection of the most reliable data from practical experience within the framework of definable limits of validity, and secondly the study of individual phenomena and mutual interactions. In this connection, the concentration limits beyond which mutual interactions come into play are of great interest.

Some processes with mutual interactions, such as comminution, agglomeration and de-agglomeration, can be quantitatively described using the methods of reaction kinetics. Here we must distinguish between the macrokinetic and microkinetic approaches. **Macrokinetics** is concerned with the course of a reaction in an apparatus; this is always influenced by the apparatus and often by individual variable factors associated with its use. **Microkinetics** describes in detail those processes which depend upon the material. The crushing of single grains and the adhesion between particles are the concern of microkinetics, whereas the course of events in a mill and in a pelletizing drum are matters for macrokinetics.

These opening remarks are intended to serve as an introduction to the outline of the whole field of mechanical process technology which now follows. In this outline the subject is not primarily divided according to the unit operations involved. The general principles are presented first, and they are followed by a description of the processes of separation, mixing, comminution and agglomeration. Basic texts relevant to this subject are those of Brauer [3], Grassmann [6], Rumpf and Schönert [20], Schubert [23] and Ullricht [28].

2
The characteristics of systems and their changes of state

2.1 DISPERSE SYSTEMS

Disperse systems generally consist of two or more phases. If the elements of a particulate material are separately embedded in a continuous phase called the **dispersion medium**, then we describe the system as **discretely disperse**, in contrast with disperse systems in which two continuous phases permeate each other. A continuous disperse phase is termed **compactly disperse**. It can also consist of elements which, although distinguishable, stick to one another or merge into one another. If, however, these elements adhere to one another without forming bridges of material, then the phase can also be termed discretely disperse. Here the designations are not clear cut.

> On sintering the disperse solid phase passes through all states: from the discretely disperse agglomerate to the continuous phase with distinguishable particles, then the continuous phase with porous inclusions (the previously continuous gaseous phase has now become discretely disperse), until finally an almost voidless solid phase is formed.

The elements of the discrete domains are called particles. The term 'grain' is also commonly used, especially in expressions such as 'grain size' and 'grain shape'. Large particles are called lumps. However, such distinctions made in everyday speech are too ill defined. We shall therefore generally refer only to particles. In the case of rock won from a quarry the particle size can be as much as several metres. Figure 2.1 shows a schematic classification of disperse systems which is subdivided first according to the phases solid (s), liquid (l) and gaseous (g) and then according to particle size.

> In the range of coarsely disperse systems there is a lack of standard designations. The terms emulsion and suspension are used in colloid chemistry for the more coarsely dispersed 'liquid-in-liquid' and 'solid-in-liquid' systems. However, we also refer to 'gas–solid suspensions' or simply gas–solid and gas–liquid mixtures. In the latter case it is often in any event

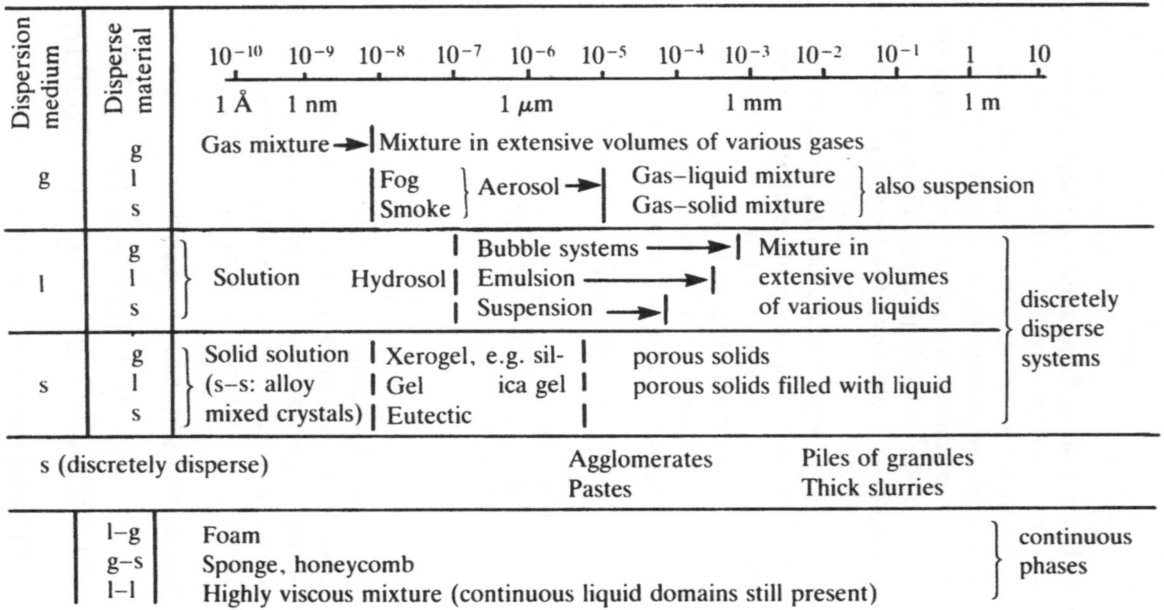

Figure 2.1 Schematic classification of disperse systems.

Dispersion medium	Disperse material	10^{-10} — 10^{-9} — 10^{-8} — 10^{-7} — 10^{-6} — 10^{-5} — 10^{-4} — 10^{-3} — 10^{-2} — 10^{-1} — 1 — 10
		1 Å — 1 nm — 1 μm — 1 mm — 1 m
g	g	Gas mixture → Mixture in extensive volumes of various gases
	l	Fog ⎱ Aerosol → Gas–liquid mixture ⎱ also suspension
	s	Smoke ⎰ Gas–solid mixture ⎰
l	g	Solution Hydrosol Bubble systems → Mixture in extensive volumes of various liquids
	l	Emulsion
	s	Suspension →
s	g	Solid solution Xerogel, e.g. sil- porous solids
	l	(s–s: alloy) Gel ica gel porous solids filled with liquid
	s	mixed crystals Eutectic

discretely disperse systems   discretely disperse systems

s (discretely disperse)   Agglomerates   Piles of granules
                      Pastes   Thick slurries

l–g   Foam
g–s   Sponge, honeycomb ⎱ continuous phases
l–l   Highly viscous mixture (continuous liquid domains still present) ⎰

difficult to distinguish exactly between the discretely disperse state and the mixing of continuous phases.

The expressions 'coarse' and 'fine' can only be used for making relative distinctions. In colloid chemistry 'coarsely disperse' systems fall in a range above 'colloidally disperse' systems, i.e. above 0.5 μm. In crushing and grinding the boundary between 'coarse milling', i.e. crushing, and 'fine milling', i.e. grinding, can lie anywhere between 10 μm and 1 mm as expressed in terms of the average particle size of the milled material.

2.2 REPRESENTING THE CHARACTERISTICS OF PARTICLE ASSEMBLIES

2.2.1 Measures of dispersity: fineness parameters

The nature of the disperse phase of a discretely disperse system has to be represented. It consists of an assembly of particles and is a population of elements (particles), which has to be described in terms of a distribution function [82]. To do this it is necessary to classify the particles according to some property, which must be a measurable physical magnitude specified by a numerical value and a unit. We call these measures of dispersity, which are directly or indirectly linked to the particle size, fineness parameters.

Fundamentally, any property by means of which the particles can be definitely classified is suitable as a measure of dispersity. Some measures of this type are preferentially used for describing particle assemblies.

(a) Geometrical measures of dispersity

Geometrical measures of dispersity are based on a geometrical quantity – a length, an area or the particle volume. With particles of irregular shape it is necessary to stipulate which particle length or area is being measured. The **main dimensions** a, b and c which are used are those which can be defined by either circumscribed or inscribed regular bodies, such as ellipsoids or cuboids, or by comparison with the largest dimension of the particle.

Statistical lengths (Fig. 2.2) are determined by an automatic scanning of pictures of the particles in which they have taken up a random position relative to the measurement. As a consequence of this variable orientation in relation to the direction of measurement there is a distribution of the statistical length for each identical non-spherical particle.

Other geometrical measures are the **volume**, the **surface area** and a **projected area of the particle** either lying in its stable position or oriented in an average random direction.

If the particles are geometrically similar, these geometric parameters are related to each other by formulae which are the same for all such particles, but in practice this is seldom the case.

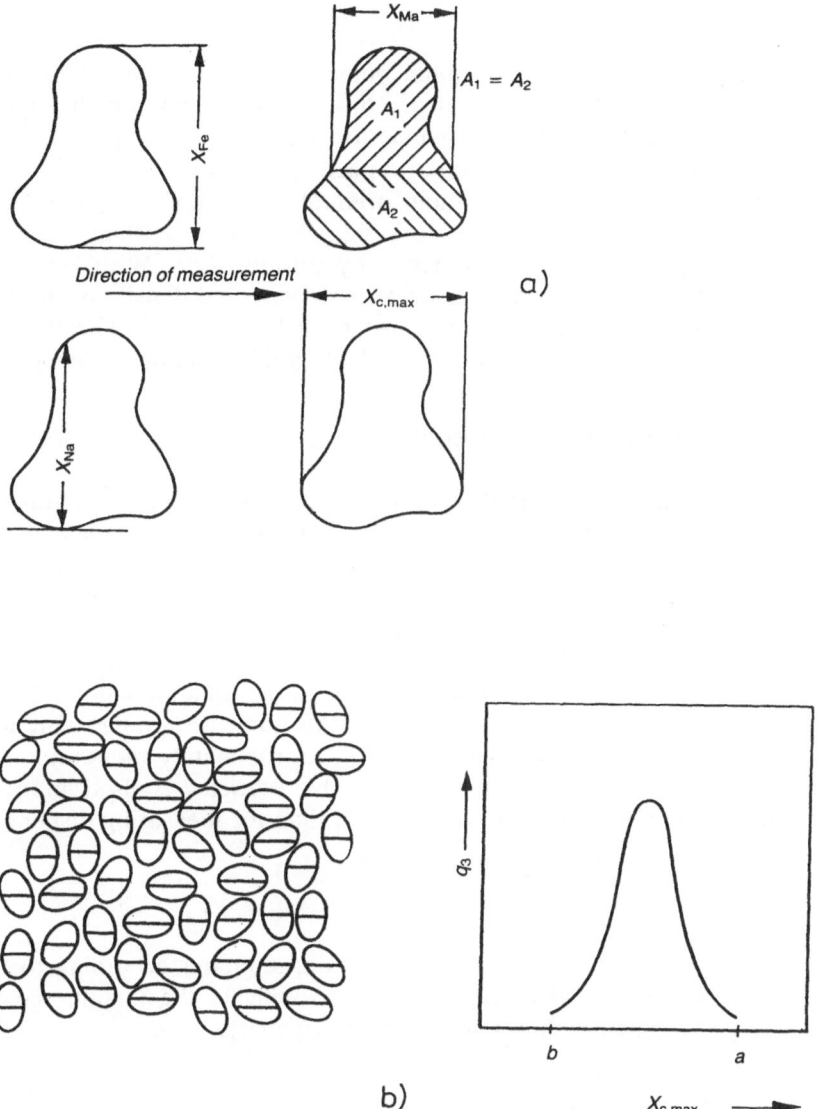

Figure 2.2 Statistical lengths. (a) Summary of the commonly used lengths: the Feret diameter x_{Fe} is the projection of the particle's outline onto a line perpendicular to the direction of measurement; the Martin diameter x_{Ma} is the chord parallel to the direction of measurement which bisects the projected area; the Nassenstein diameter x_{Na} is the chord perpendicular at the point of contact to a tangent parallel to the direction of measurement; the diameter $x_{c,max}$ is the longest chord parallel to the direction of measurement. (b) Distribution of a statistical length for a narrow range of particle size.

(b) Settling velocity

The settling velocity is the terminal velocity of a particle falling under gravity in an infinite expanse of a medium at rest. This quantity can be measured, and it has a definite value for spheres and roughly spherical (isometric) particles. For non-isometric particles the settling velocity in the Stokes law range depends on their orientation to the direction of the gravitational force. We then have to assign an average settling velocity to the particle. It is measured in practice by settling tests. With Newtonian fluids the settling velocity depends on the viscosity and density of the fluid, so that its nature and temperature must both be specified. The fluid and particle mechanics of sedimentation are dealt with at length in Chapter 3.

(c) Optical measures of dispersity

If we wish to measure particles without interfering with them during a process (e.g. if a cloud of particles is passing through a duct) then the best method available is the light-scattering procedure. In this method the intensity of the scattered light emitted from an illuminated particle at a given angle to the direction of illumination (the angle of scatter) is a measure of dispersity. With spheres of diameters greater than 5 μm and less than about 0.5 μm this intensity is a monotonic function of their diameter. With non-spherical particles in this same size range and for a given angle of scatter the relation remains approximately monotonic. The extinction of a beam of light or X-rays by particles is used more for determining the cumulative size distribution than for measuring dispersity.

(d) Density

In ore dressing we are interested in separating minerals according to their chemical composition. The density serves to indicate the metallic content of an ore particle and is then a measure of dispersity.

2.2.2 Equivalent diameter

Many properties of particles which are used as measures of dispersity can be expressed in terms of equivalent diameters (Table 2.1). These are the dimensions of a particle of defined geometrical shape which has the same property as the property of interest of the particle under investigation. The equivalent diameters are then introduced as measures of dispersity; for instance, the equivalent settling rate or drag diameter is the diameter of a sphere with the same terminal velocity and density as the particle when it

Table 2.1 The most important equivalent diameters

Geometric equivalent diameters		
Diameter of the sphere with the same volume		x_v
Diameter of the sphere with the same surface area		x_s
Diameter of the sphere with the same specific surface area		x_{sv}
Diameter of the circumscribing sphere		x_{en}
From the projection of the particle	Particle orientation	
	Stable	Average
Diameter of the circle with the same area	$x_{p,s}$	$x_{p,m}$
Diameter of the circle with the same perimeter	$x_{pe,s}$	$x_{pe,m}$
Diameter of the inscribed circle	$x_{in,s}$	$x_{in,m}$
Diameter of the circumscribed circle	$x_{ci,s}$	$x_{ci,m}$
Hydrodynamic equivalent diameters		
Diameter of the sphere with the same resistance (drag)		x_D
Diameter of the sphere with the same terminal velocity		x_u
Diameter of the sphere with the same terminal velocity in the Stokes law range (Stokes diameter)		x_{St}
Other equivalent diameters		
Diameter of the sphere scattering light at the same intensity		x_{sca}
Diameter of the sphere causing the same change in electrical resistance (Coulter counter)		x_{el}

is falling in the same fluid. If the velocity of the particle is measured in the Stokes law region, then the drag diameter is called the Stokes diameter. If the projected area of a particle is compared with a circle of the same area, then the equivalent diameter is that of the circle corresponding to the particle lying in either its stable or its average position. The introduction of an equivalent diameter proves to be very appropriate in many cases, particularly when, for instance, a linear measurement must be converted into an area and a volume.

If we are interested in the motion of a particle in a fluid then the appropriate measure of dispersity is the settling or terminal velocity. It is usually measured in the Stokes law region. The Stokes diameter thus obtained is truly characteristic of the behaviour of the particle, provided also that the particle is actually moving in the Stokes law range. If the motion occurs at higher Reynolds numbers (see Chapter 3, section 3.1.1(b)), we generally again use the Stokes diameter to specify it, together with the laws for the drag force in the intermediate and Newtonian ranges. However, this is not exact. Strictly speaking that equivalent diameter which corresponds to the effective particle resistance should be chosen. It will probably deviate from the Stokes diameter because at higher Reynolds numbers drag plays a more important role than in the Stokes law range. We could also consider using the equivalent diameter of the circle with the same projected area; however, it is not certain that this would reduce the error. Using the carefully measured Stokes diameter is often the most suitable approximation. Frequently an equivalent diameter is used which has been obtained by sifting. If this is done even larger errors can arise.

2.2.3 Particle shape

We distinguish between the size of a particle and its shape. Spheres have the same shape but may be of a different size. With irregularly formed or broken particles the particle shape could also, in principle, be used as a measure of dispersity since, strictly speaking, each particle is distinguished from the others by its shape. Certain particle shapes are preferred in many applications, e.g. isometric particles for road metal or sharp-edged grains for abrasives. However, a differentiation on the basis of shape alone is rarely of interest. As a general rule, even in the examples mentioned above, the shape criteria are used together with other properties, mainly particle size. In the milling of cereals, for instance, the separation of the grains of meal from the bran is important, as is the separation of the peeled grains from the husks and unpeeled grain in the case of oats and rice. Here the behaviour of the particles being separated is governed by their shape, density, surface roughness and size.

Particle shape is thus a very important property and influences all those properties put forward above as measures of dispersity. If geometrically defined dimensions or equivalent diameters are used as measures of dispersity, then the influence of shape is allowed for by means of a **shape factor** $\psi_{\alpha,\beta}$ which is defined as follows [72]:

$$x_\alpha = \psi_{\alpha,\beta} x_\beta \tag{2.1}$$

where x_α and x_β stand for any two geometric measures of dispersity or equivalent diameters with appropriate subscripts α and β. If, for instance, we write $x_s = \psi_{s,v} x_v$, then $\psi_{s,v}$ is the factor for converting the diameter x_v of the sphere with the same volume, i.e. the volume diameter, into the diameter x_s of the sphere with the same surface area, i.e. the surface diameter. The relation $x_s \geqslant x_v$ always applies, where the equals sign is valid only for spherical particles. For all other particles the surface area of the sphere of the same volume is less than the actual surface area; in other words, the sphere with the same surface area is larger than the sphere with the same volume. The parameter

$$\psi_{v,s}{}^2 = \frac{x_v{}^2}{x_s{}^2} \leqslant 1 \tag{2.2}$$

is called the sphericity [98]. According to a theorem due to Cauchy the following relation holds for convex particles:

$$x_s = x_{pm} \quad \text{or} \quad \psi_{s,pm} = 1 \tag{2.3}$$

A series of measured shape factors is given in Table 2.2.

Table 2.2 Measured shape factors

Material		$\psi^* = \dfrac{(\pi/6)x_v{}^3}{x_{p,s}{}^3} = \dfrac{\pi}{6}\psi_{v,ps}{}^3$	Sphericity $\psi_{v,s}{}^2$ (estimated from $\psi_{v,ps}$)
Coal			
Anthracite	20 μm	0.20–0.28	0.5–0.7
Natural coal dust	124 μm	0.20	0.65
Coal	124 μm	0.25	0.73
Coal	2411 μm	0.25	0.75
Sand			
Pulverized	40 μm	0.14	0.55
		–	0.4–0.18
Crushed	40 μm to 3 mm	0.15	0.53
		0.17–0.28	0.6–0.75
			0.8–0.9
Rounded-off grain	40 μm to 3 mm	0.34	0.8
Air-borne sand	40 μm to 3 mm	–	0.95
Elutriated sand	40 μm to 3 mm	–	0.92–0.98
Various materials			
Fly-ash (ball-shaped)	124 μm	0.41	0.89
Tungsten	124 μm	0.45	0.89
Glass (pulverized)	124 μm	0.28	0.65
Fibrous black coal	124 μm	0.10	0.38
Mica flakes	124 μm	0.03	0.28
Sillimanite	2411 μm	0.25	0.75
Cement	40 μm	–	0.7–0.8

2.2.4 General representation of particle assemblies

(a) Definition of distribution functions for various measures of populations

Let x be any measure of dispersity. The population of particles being measured is distributed according to some function of x. It is appropriate to choose x to be a linear dimension such as the equivalent diameter. The numerical value of this dimension can, in principle, lie anywhere between zero and infinity. In fact, however, there is always a minimum value x_{min} and a maximum value x_{max}. The distribution function $Q_r(x)$ indicates which proportion of the total population lies between x_{min} and x. Thus (cf. Fig. 2.3):

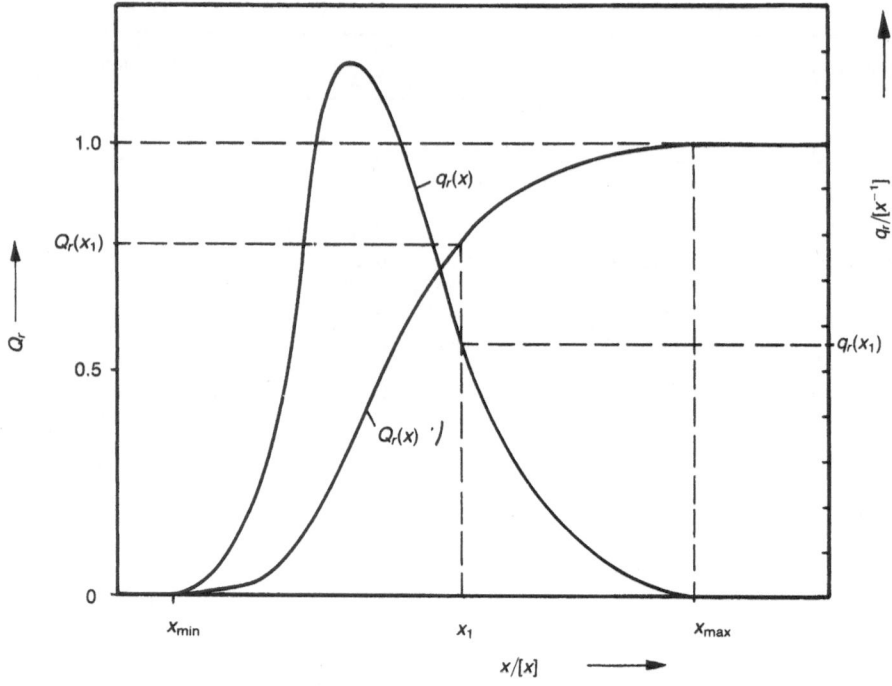

Figure 2.3 Graphical representation of particle assemblies: $q_r(x)$, density distribution; $Q_r(x)$, cumulative function distribution; $[x]$, measure of dispersity.

$$0 < Q_r(x) < 1 \qquad\qquad\qquad\qquad\qquad\qquad\qquad (2.4)$$

$$Q_r(x_{min}) = 0 \quad\text{and}\quad Q_r(x_{max}) = 1 \qquad\qquad\qquad (2.4a)$$

In introducing $Q_r(x)$ the question as to how the population is to be measured has been left open; it can, for example, be measured by number, by volume, by mass etc. The subscript r is used to designate the kind of measure being used. For example, if the particles are counted, then the population is measured by number and we write $Q_0(x)$, by length $Q_1(x)$, by surface area $Q_2(x)$ and by volume $Q_3(x)$. For distributions by mass we shall retain the traditional designations of calling the fraction undersize the complement C and the fraction oversize the residue R, where $C(x) + R(x) = 1$. If the density is independent of the particle size then $C(x) = Q_3(x)$.

With most of the particle assemblies which occur in practice the number of particles is so large that the distribution function $Q_r(x)$ can be regarded as continuous and it can be differentiated as follows:

$$q_r(x) = \frac{dQ_r(x)}{dx} \qquad \text{in particular } y(x) = \frac{dC(x)}{dx} \qquad (2.5)$$

where $q_r(x)$ is the **density distribution function** corresponding to the cumulative distribution function $Q_r(x)$. In the case of $Q_0(x)$, the cumulative distribution by number $q_0(x)$ is the **frequency distribution**. Similarly $y(x)$ and $q_3(x)$ are the **density distributions by mass and volume**.

Conversely, the cumulative distribution function can be obtained from the population density by integration:

$$Q_r(x) = \int_{x_{min}}^{x} q_r(x)\,dx \qquad (2.6)$$

Thus $Q_r(x)$ is also known as the cumulative distribution of the population; particular examples of $Q_r(x)$ are the cumulative distribution by number $Q_0(x)$ and the cumulative distribution by mass $C(x)$. However, it is appropriate, in accord with the definitions used in the theory of statistics, to refer to the cumulative distribution simply as the distribution function.

The concept of 'density' signifies the differential increment in the population corresponding to a given differential increment in the measure of dispersity. In a similar way in the theory of statistics the probability is represented as a distribution function of a random variable whose derivative is the probability density.

(b) Moments

For calculations involving distributions, condensed expressions called **moments** have been introduced for certain integrals. It is well known that these formal expressions cause the novice certain difficulties, although, once he has gained some familiarity with them, they are very simple to use and greatly facilitate the mathematical manipulation of distributions. (Formal definitions are like languages which can only be used to advantage when it is not necessary constantly to consult a dictionary.) We define the moments as follows:

kth incomplete moment of the Q_r distribution between the limits x_1 and x_2:

$$M_{k,r}(x_1, x_2) = \int_{x_1}^{x_2} x^k q_r(x)\,dx \qquad (2.7a)$$

kth complete moment of the Q_r distribution:

$$M_{k,r} = M_{k,r}(x_{min}, x_{max}) = \int_{x_{min}}^{x_{max}} x^k q_r(x)\,dx \qquad (2.7b)$$

The complete moment $M_{k,0}$ is the integral average value of the term x^k weighted according to the frequency distribution $q_0(x)$:

$$\overline{(x^k)} = M_{k,0} = \int_{x_{min}}^{x_{max}} x^k q_0(x)\,dx \qquad (2.7c)$$

(c) Conversion of distributions

It is often necessary to convert the distribution of a population measured in one way into the distribution measured in another way. For example, given the distribution by number (frequency distribution) how does one find the distribution by mass? For the general case, where the new measure of the population is proportional to the rth power of x, the following relation holds:

$$q_r(x) = \frac{x^r q_0(x)}{\int_{x_{min}}^{x_{max}} x^r q_0(x)\,dx} = \frac{x^r q_0(x)}{M_{r,0}} \tag{2.8}$$

where $q_0(x)$ is the frequency distribution of particles characterized by some measure of dispersity x. When carrying out conversions of distributions a suitable measure of dispersity to choose is a linear dimension such as the equivalent diameter.

(d) Specific statistical values of the measure of dispersity

As shown in Fig. 2.4 certain specific x values are characteristics of the distribution function $Q_r(x)$. There is the median value $x_{50,r}$ for which the defining relation is $Q_r(x_{50,r}) = 0.5$. This designates the x value corresponding to $Q_r = 0.5$. The quartiles are defined by the relations $Q_r(x_{25,r}) = 0.25$ and $Q_r(x_{75,r}) = 0.75$ respectively. The general relation is

$$Q_r(x_{a,r}) = a/100 \tag{2.9}$$

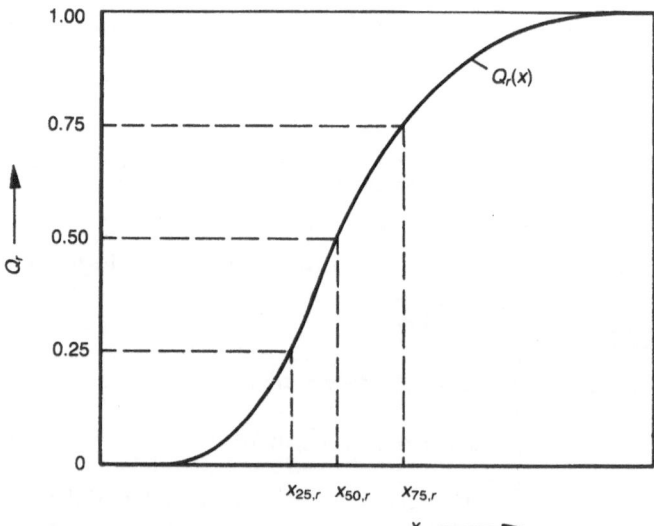

Figure 2.4 Specific x values of a distribution function.

(e) The spread of a distribution: measures of scatter

The whole spread of a distribution is specified by the difference $x_{max} - x_{min}$ or by the ratio x_{max}/x_{min} or by the logarithm of this ratio, i.e. $\log(x_{max}/x_{min}) = \log x_{max} - \log x_{min}$. Near x_{max} and x_{min} the distribution functions frequently approach the horizontal asymptotically and, as a consequence, x_{max} and x_{min} are very difficult to measure. The spread of the distribution may be more exactly fixed by pairs of values such as $x_{99.9}/x_{0.1}$, x_{99}/x_1, x_{90}/x_{10}, x_{75}/x_{25} etc.

Mathematical functions for representing $Q_r(x)$ are given in section 2.2.6(b). They contain parameters which are a measure of the spread of the distribution. For any given function $Q_r(x)$ the statistical measures of scatter, the variance σ_r^2 and the standard deviation σ_r, are measures of the spread of the distribution defined by the integral

$$\sigma_r^2 = \int_{x_{min}}^{x_{max}} (x - \bar{x}_r)^2 q_r(x)\, dx = M_{2,r} - M_{1,r}^2 \tag{2.10}$$

(f) Specific surface

If the surface area of a particle or a collection of identical particles of size x is designated by $S_e(x)$, its volume by $V_e(x)$ and its mass by $M_e(x)$, then the volume- and mass-related specific surfaces of the particle(s) are

$$S_{v,e}(x) = \frac{S_e(x)}{V_e(x)} \quad \text{and} \quad S_{m,e}(x) = \frac{S_e(x)}{M_e(x)} \tag{2.11}$$

The surface area $S_e(x)$ can be defined in very different ways. The definition depends upon the method of measurement used. If the surface area is measured using gas adsorption, a 'gas adsorption' surface is obtained which accounts for the effects of molecular roughness and the distribution of energy over the surface. If a permeability method is used, then only those roughnesses (asperities) are registered that affect the resistance to flow. The determination of a surface area by means of an extinction method leads to an 'extinction surface'. The theoretical and technical problems associated with these methods of measurement will not be discussed here. The surface area can also be calculated when the shape and, if necessary, the roughness of the particle are known.

If it is assumed that the surface area $S_e(x)$ of a particle can be defined, then an equivalent diameter x_s can be derived from the formula

$$S_e(x) = \pi x_s^2 = \pi \psi_{s,x}^2 x^2 \tag{2.12}$$

The shape factor $\psi_{s,x}$ transforms the given measure of dispersity x into the surface diameter x_s. Following the same rule we can write $V_e(x) = (\pi/6)\psi_{v,x}^3 x^3$ for the volume $V_e(x)$ of a single grain and $M_e(x) = \rho(x)V_e(x)$ for its mass. Thus the specific surface of a particle is given by

$$S_{v,e}(x) = 6\frac{x_s^2}{x_v^3} = 6\psi_{s,v}^2 x_v^{-1} = 6\frac{\psi_{s,x}^2}{\psi_{v,x}^3} x^{-1} \tag{2.13a}$$

$$S_{m,e}(x) = \frac{S_{v,e}(x)}{\rho(x)} \tag{2.13b}$$

The corresponding relation for the specific surface of the whole collection of particles is

$$S_v = \int_{x_{min}}^{x_{max}} S_e(x) q_0(x)\, dx \Big/ \int_{x_{min}}^{x_{max}} V_e(x) q_0(x)\, dx \tag{2.14a}$$

i.e.

$$S_v = 6 \int_{x_{min}}^{x_{max}} \psi_{s,x}^2 x^2 q_0(x)\, dx \Big/ \int_{x_{min}}^{x_{max}} \psi_{v,x}^3 x^3 q_0(x)\, dx \tag{2.14b}$$

In the case of geometrically similar particles $\psi_{s,x}$, $\psi_{v,x}$ and $\psi_{s,v}$ are constant. It then follows that

$$S_v = 6 \frac{\psi_{s,x}^2}{\psi_{v,x}^3} \frac{M_{2,0}}{M_{3,0}} = 6 \frac{\psi_{s,v}^2}{\psi_{v,x}} \frac{M_{2,0}}{M_{3,0}} \tag{2.14c}$$

If only the specific surface of that fraction of a particle assembly lying between x_1 and x_2 is to be considered, then it follows by analogy that

$$S_v(x_1, x_2) = 6 \frac{\psi_{s,x}^2}{\psi_{v,x}^3} \frac{M_{2,0}(x_1, x_2)}{M_{3,0}(x_1, x_2)} \tag{2.15a}$$

Since the distribution by volume $Q_3(x)$ is frequently measured, the above equation can be transformed into the more convenient form

$$S_v(x_1, x_2) = 6 \frac{\psi_{s,x}^2}{\psi_{v,s}^3} \frac{M_{-1,3}(x_1, x_2)}{Q_3(x_2) - Q_3(x_1)} \tag{2.15b}$$

In general the particles are not geometrically similar. Despite this, calculations are performed with constant shape factors which are then interpreted as average factors.

When the density $\rho(x)$ is constant, the relation $S_m = S_v/\rho$ always holds. Instead of the distribution function by volume $Q_3(x)$ and $M_{-1,3}$, the corresponding terms $C(x)$ and

$$M_{-1,y}(x_1, x_2) = \int_{x_1}^{x_2} \frac{y(x)}{x}\, dx$$

are introduced into the equation.

2.2.5 The Gaussian or normal distribution

The normal distribution, which is due to Gauss, is the most important distribution in statistics and, as such, is also of importance for mechanical process technology. We make use of it mainly for statistical problems such as occur with mixing for example. The density distribution is symmetrical about the mean and therefore would not be expected to apply to the size distribution of particle assemblies except for very narrow distributions.

However, the logarithmic normal (log-normal) distribution, whose density distribution is a symmetrical function of the logarithm of the variate, is suitable for representing grain size distributions, as demonstrated in section 2.2.6(b). It is therefore pertinent at this stage to introduce the normal distribution.

The cumulative and density distribution functions are

$$H(x) = \int_{-\infty}^{x} h(x)\,dx$$

$$h(x) = \{\sigma_x \sqrt{(2\pi)}\}^{-1} \exp\left\{-\frac{(x-\bar{x})^2}{2\sigma_x^2}\right\} \qquad (2.16)$$

where

$$\bar{x} = \int_{-\infty}^{+\infty} xh(x)\,dx \qquad \sigma_x^2 = \int_{-\infty}^{+\infty} (x-\bar{x})^2 h(x)\,dx$$

If t is substituted for $(x-\bar{x})/\sigma_x$, then the standardized normal distribution is obtained with mean $t = 0$ and variance $\sigma_t^2 = 1$:

$$H(t) = \int_{-\infty}^{t} h(t)\,dt \qquad \text{where} \qquad h(t) = \{\sqrt{(2\pi)}\}^{-1} \exp\left(\frac{-t^2}{2}\right) \qquad (2.17)$$

The values of the standardized distribution function $H(t)$, or the values of $H(t) - H(0) = H(t) - 0.5$, are tabulated and can be found in any textbook of statistical mathematics.

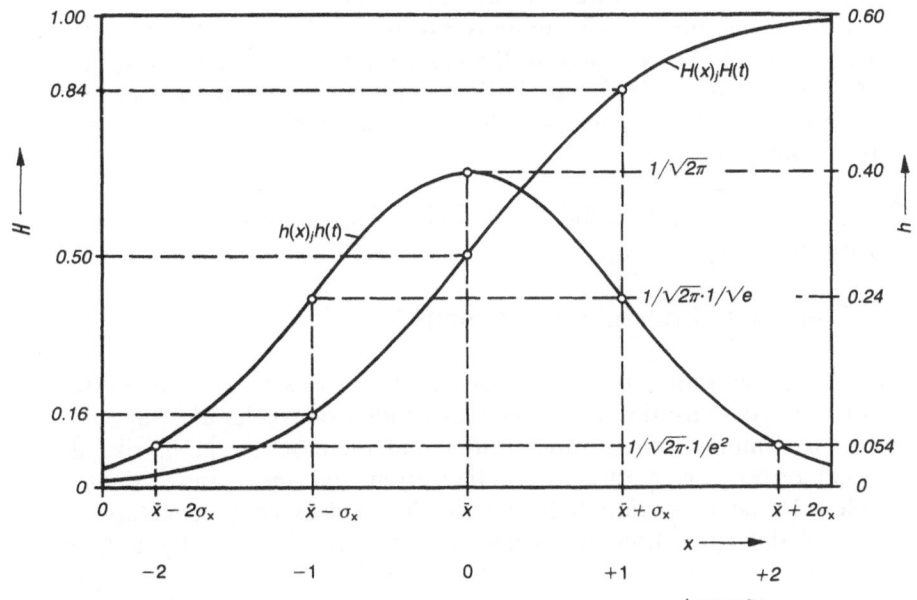

Figure 2.5 The Gaussian or normal distribution.

Table 2.3 Some important ordinate values of the standardized normal distribution

t	$H(-t)$	$H(+t)$	$H(t) - H(-t)$
0	0.5000	0.5000	0.0000
0.5	0.3085	0.6915	0.3829
1.0	0.1587	0.8413	0.6827
1.5	0.0668	0.9332	0.8664
2.0	0.0228	0.9772	0.9545
2.5	0.0062	0.9938	0.9876
3.0	0.0013	0.9987	0.9973

The bell-shaped curve $h(t)$ and the cumulative curve $H(t)$ are plotted against x and t as abscissae in Fig. 2.5. The points of inflexion of the bell-shaped curve occur at $x = \bar{x} \pm \sigma_x$, i.e. at $t = \pm 1$. Some important ordinate values are given in Table 2.3.

Because of the symmetry of the normal distribution it follows that

$$H(+t) - H(-t) = 1 - 2H(-t) \tag{2.18}$$

The normal distribution is of great practical importance because it is a consequence of the addition of independent random events. This concept was established by Gauss in his hypothesis of primary errors. If an error is the sum of n independent measurements, each of which possesses a small scatter in comparison with the scatter of the sum, then its distribution approaches the normal distribution asymptotically, i.e. for $n \to \infty$. The same statement is formulated in the central limit theorem of statistics. The normal distribution $H(t)$ can be represented by a straight line if it is plotted on probability paper which has an ordinate scale with graduations corresponding to $H(t)$.

2.2.6 Particle distributions and distribution functions approximating to them

(a) Examples of particle distributions

Particle distributions arise and undergo changes as a result of crystallization or the condensation of amorphous solids from melts and vapours, as a result of comminution, agglomeration and increases in grain size due to thermal effects, as a result of nebulization and the formation of liquid droplets by condensation followed by their subsequent coagulation, as a result of the formation and coalescence of gas bubbles by aeration or evaporation, or as a result of a whole number of mixing and separating processes. Therefore it is not to be expected that there would be a uniform law for the formation of particle distributions. A certain uniformity will come about because, in the formation of distributions, there is often an interplay between a large number of chance factors. This would favour the emergence of normal distributions. However, they occur only in the case of narrow distributions. If the chance events act in a multiplicative rather than an additive way, then the log-normal distribution follows from the theory. In fact it is frequently observed that particle assemblies can be approximately described by log-normal distributions.

Comminution, agglomeration and crystallization do not produce equi-sized grains, but rather a distribution of sizes. The nature of the distribution is determined not only by independent random processes but also, and often substantially, by systematic influences. Furthermore, selective influences and limiting conditions restrict the free play of random factors. With comminution the size of the largest particle is determined by the size of the particles in the feed, just as there is a lower limit of grindability, and hence the distributions have relatively sharp limits (see Fig. 2.6(a)). As soon as the distributions are substantially influenced by separation and mixing their shape can become irregular. When two distributions with widely separated means x_a and x_b are mixed together, bimodal distributions arise whose density functions show two maxima (Fig. 2.6(b)). If we encounter such a distribution without knowing how it has arisen, then the lack of material between the sizes x_a and x_b can be ascribed to misplaced particles. The coarse and fine fractions of a sharp size grading only slightly overlap and the plots of their (density) distributions have steeply sloping sides. At the fine end of the distribution, i.e. at particle sizes of about 10 μm, agglomeration can lead to the formation of clusters of the finest particles and hence result in a corresponding reduction in the measured particle distribution (Fig. 2.6(c)).

(b) The log-normal distribution

If we apply the central limit theorem to a product of independent random measurements $x = (z_1 z_2 \ldots z_n)$, then $\log x = (\log z_1 + \log z_2 + \ldots + \log z_n)$ is a sum of random measurements. The magnitude $\log x$ is thus asymptotically normally distributed and the magnitude x is asymptotically log-normally distributed. Let us assume that a measure of dispersity x is log-normally distributed. Then

$$Q_r(x) = H(t) \quad \text{and} \quad q_r(x)\,\mathrm{d}x = h(t)\,\mathrm{d}t \tag{2.19}$$

where the following substitutions apply:

$$t = \frac{\ln x - \ln x_{50,r}}{\sigma} = \frac{\ln (x/x_{50,r})}{\sigma} \qquad x = x_{50,r} \exp(\sigma t) \tag{2.19a}$$

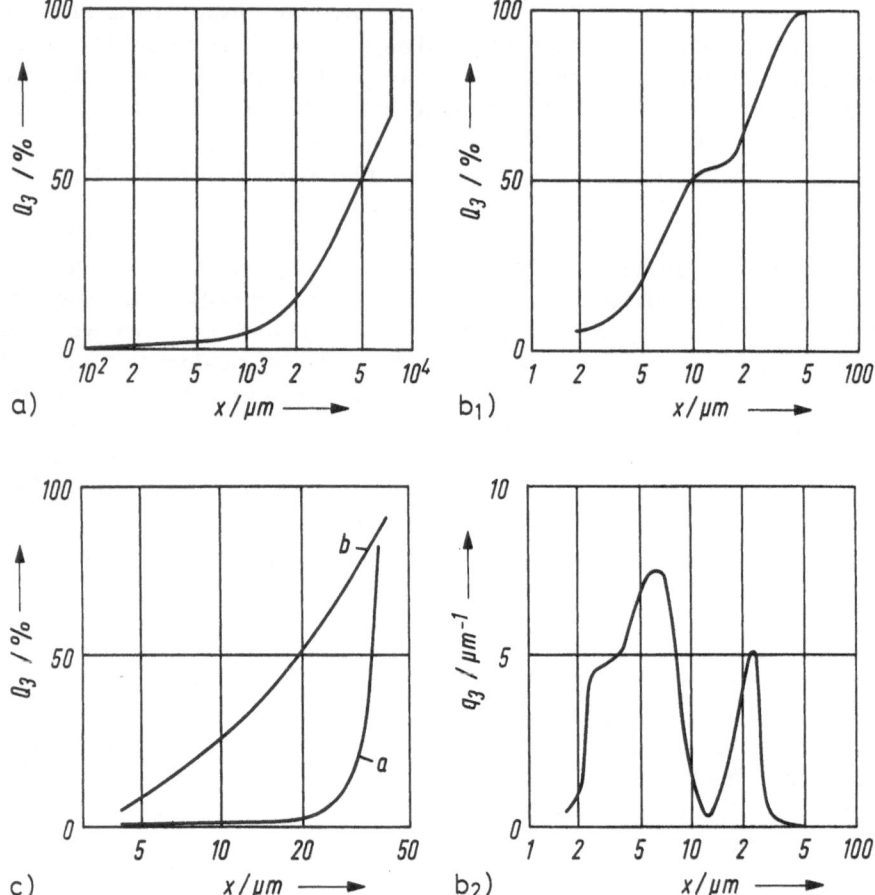

Figure 2.6 Examples of particle distributions (special cases): (a) impact breakage of glass spheres; (b) bimodal distributions of an assembly of limestone particles ((b_1) distribution function, (b_2) density distribution); (c) sedimentation analysis of an agglomerated limestone fraction (curve a) and the same fraction after de-agglomeration (curve b).

σ, which is dimensionless, is the standard deviation of the log-normal distribution. It corresponds on the x scale to the value σ_x according to the expression $\sigma = \ln \sigma_x - \ln x_{50,r}$, for $x = \sigma_x$, $t = 1$. The following conditions hold for x and t: $x = x_{min} = 0$ corresponds to $t = -\infty$, $x = x_{max} = \infty$ corresponds to $t = \infty$ and for $x = x_{50,r}$, $t = 0$. The boundary conditions $x_{min} = 0$ and $x_{max} = \infty$ are not realizable. Despite this the log-normal distribution, particularly in the fine region, is often well suited to the approximate representation of actual distributions (Fig. 2.7). The actual $x_{min} > 0$ is generally not measurable and the condition $x_{max} = \infty$ limits the

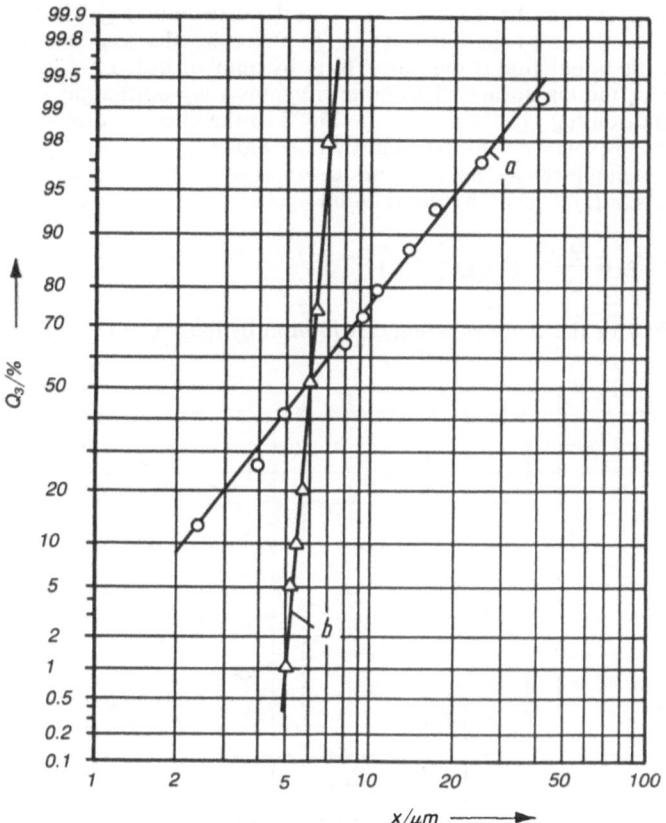

Figure 2.7 Log-normal distributions plotted on probability graph paper: curve a, size analysis of limestone (Andreasen pipette method); curve b, paraffin (kerosene) aerosol.

applicability of the log-normal distribution at the coarse end of the range:

$$M_{k,r}(0, x_i) = (x_{50,r})^k \exp\left(\frac{k^2\sigma^2}{2}\right) H(t_i - k\sigma) \qquad (2.20a)$$

$$M_{k,r} = (x_{50,r})^k \exp\left(\frac{k^2\sigma^2}{2}\right) \qquad (2.20b)$$

A particular advantage of the log-normal distribution is that all complete moments are finite. A distribution Q_r can be transformed into any other distribution Q_s by the relation between the moments:

$$Q_s(x) = \frac{M_{s-r,r}(0, x)}{M_{s-r,r}} = H\{t - (s - r)\sigma\} \qquad (2.21)$$

In changing from the distribution Q_r to the distribution Q_s the H distribution is thus displaced by the length $(s - r)\sigma$ along the abscissa. When the way in which the population is measured (e.g. by number instead of by mass) is changed, the log-normal distribution remains a log-normal distribution, and on a plot on log-probability paper (Fig. 2.8) the change corresponds to a parallel displacement to the right by $(s - r)\sigma$.

The volume-related specific surface for a given distribution by number $Q_0(x) = H(t)$ is given by the relation

$$S_v = 6 \frac{{\psi_{s,x}}^2}{{\psi_{v,x}}^3} (x_{50.0})^{-1} \exp(-2.5\sigma^2) \qquad (2.22a)$$

and that for the corresponding distribution by mass $Q_3(x) = H(t)$ is given by

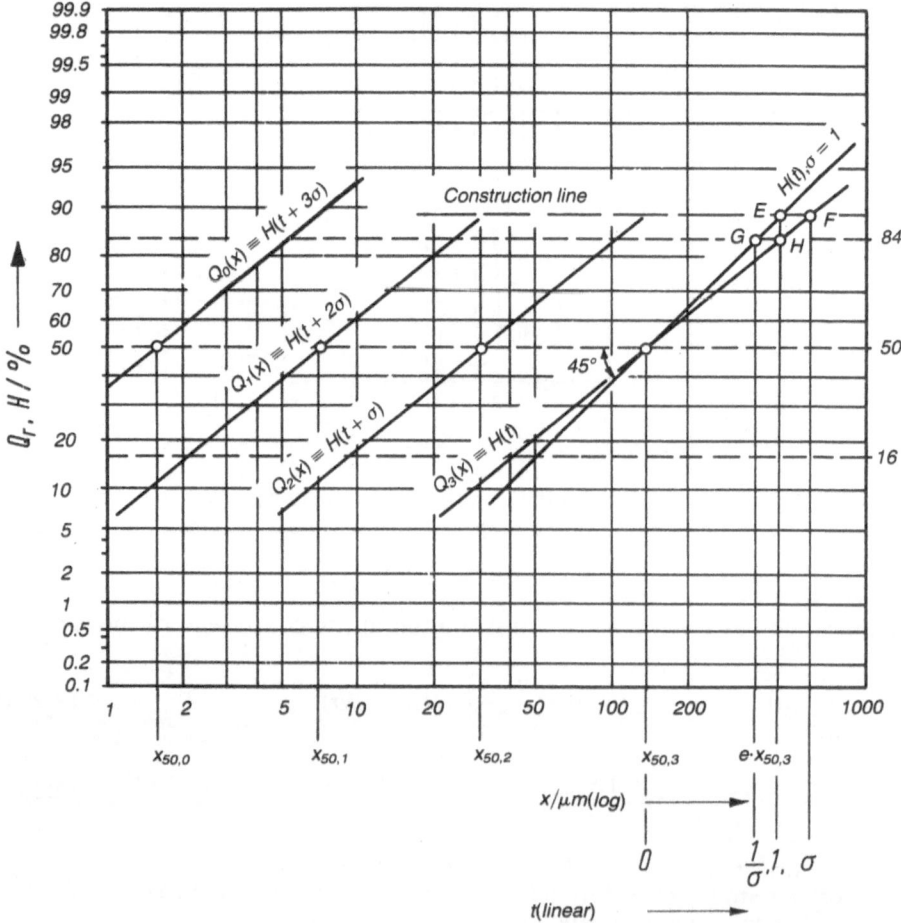

Figure 2.8 Changing the measure of population for a log-normal distribution and the construction for representing the standard deviation on log-probability paper.

$$S_v = 6 \frac{\psi_{s.x}^2}{\psi_{v.x}^3} (x_{50.3})^{-1} \exp(0.5\sigma^2) \tag{2.22b}$$

On log-probability paper (Fig. 2.8) the distribution $Q_r(x) = H(t)$ is represented by a straight line. x is plotted along the abscissa on a logarithmic scale. At $H = 0.5$ the median value $x_{50.r}$, e.g. $x_{50.3}$, is read off; the standard deviation σ is obtained from the intersection of the linear plot of $Q_r(x)$ (represented here by $Q_3(x)$, but it could also be $Q_0(x)$, $Q_1(x)$ or $Q_2(x)$) with the construction line drawn as shown in Fig. 2.8. The log-probability paper is constructed such that the plot of $H(t)$ for $\sigma = 1$ has a gradient of 45°. If σ is not equal to 1, we can find two values from the t scale: for $x = ex_{50.r}$, $t = 1/\sigma$ (point G) and for $H = 0.84$, $t = 1$ (point H). The value $1/\sigma$ must lie below the point of intersection of the line $H = 0.84$ with the 45° line, because for the latter line $\sigma = 1$. From point H the point of intersection E with the 45° line is reached and thus the construction line is obtained as shown. The point of intersection F of the construction line with the distribution function $H(t)$ then gives the distance along the abscissa from the median value that is equal to the value of σ. The larger σ is, the flatter is the plot of $H(t)$. Thus σ is a measure of the spread of the distribution, while $x_{50.r}$ is its location parameter. With some commercially available log-probability papers the values of σ and $\exp(0.5\sigma^2)$ can be read off from scales along the margins.

(c) The Gaudin–Schuhmann or power-law distribution function

Gaudin and Schuhmann have proposed a simple power law for representing the cumulative distribution by mass:

$$C(x) = \left(\frac{x}{x_{max}}\right)^m \qquad \text{for } x \leqslant x_{max}$$

$$C(x) = 1 \qquad \text{for } x = x_{max} \tag{2.23}$$

On log–log paper this distribution is plotted as a straight line. The distribution parameter m is the gradient of this line and the location parameter x_{max} is the x value for which $C = 1$. At this point the plot of the distribution suddenly becomes horizontal.

Measured distributions have no sharp bends. Even when they can be represented over the central range by a power law they deviate from this at $C \approx 0.8$–0.9 and often approach the horizontal asymptotically. Below $C = 0.8$ the gradient of milled materials lies in the range $0.5 < m < 2$.

The advantage of the Gaudin–Schuhmann distribution is that it can be simply represented on ordinary log–log paper; however, measured size distributions deviate considerably from this approximation not only in the coarse range but also in the fine range. In the latter range m must be greater than 3 on physical grounds because otherwise the analysis of this function predicts an infinitely large number of particles, which can readily be shown by means of the moments. Since every measured distribution can be approximately represented on log–log paper by a series of chords and the moments can then easily be worked out, the Gaudin–Schuhmann distribution, applied stepwise, is very useful not only for representing measured particle distributions but particularly for all calculations using these distributions. The distribution is very frequently used in the American technical literature, particularly in the field of ore dressing.

(d) The RRSB distribution

When analysing the size distributions of pulverized coal Rosin, Rammler, Sperling and Bennett (RRSB) found, more or less independently of each other, a function which was a very good approximation for representing the whole of the measured distribution. It has the form

$$R(x) = 1 - C(x) = \exp\left\{-\left(\frac{x}{x'}\right)^n\right\} \tag{2.24}$$

Like the two approximating functions dealt with above, the RRSB distribution has two parameters. The location parameter x' is defined by the relation $R(x') = 1/e = 0.368$. The exponent n is the parameter giving the spread of the distribution. The series expansion in terms of $(x/x')^n$ shows that, for small values of x/x', the RRSB distribution becomes the same as the Gaudin–Schuhmann distribution. For ground materials the exponent n also falls approximately in the same range of $0.5 \leqslant n \leqslant 2$ as m does in the Gaudin–Schuhmann distribution. In the coarse range measured distributions are often reproduced with surprising accuracy. For this reason the RRSB distribution has been very well accepted in practice. Special Rosin–Rammler paper prepared according to German Standard DIN 4190 is available for plotting it; the abscissa of this paper has a log scale and the ordinate has a log–log scale.

The linear form of the RRSB equation is $\log\log(1/R) = n\log x - n\log x' + \log\log e$ and a plot of $\log\log(1/R)$ against $\log x$ thus gives a straight line. The ordinate value $C = 1$ is first reached at $x \to \infty$. The distribution function thus has no salient point (sharp bend). In the fine region the same problems arise as with the Gaudin–Schuhmann distribution: for values of $n < 3$, $M_{-3,3}$, and hence the number of particles, becomes infinite. The moments lead to gamma functions which, although they are tabulated, are more awkward to manipulate than the integrals of the log-normal and Gaudin–Schuhmann distributions.

2.2.7 Similar distributions

We call various distribution functions $Q_r^{(1)}(x)$, $Q_r^{(2)}(x)$, $Q_r^{(3)}(x)$, ..., $Q_r^{(i)}(x)$ similar if, when the dimensionless ratio $x/x_{(i)}{}^*$ is substituted for x, a single identical distribution function is obtained. The reference size $x_{(i)}{}^*$ can be any specific statistical value of the measure of dispersity which is defined in the same way for each distribution, e.g. $x_{50,r}$, $x_{a,r}$, x_{max}, x_{min}, $\bar{x}_r$.

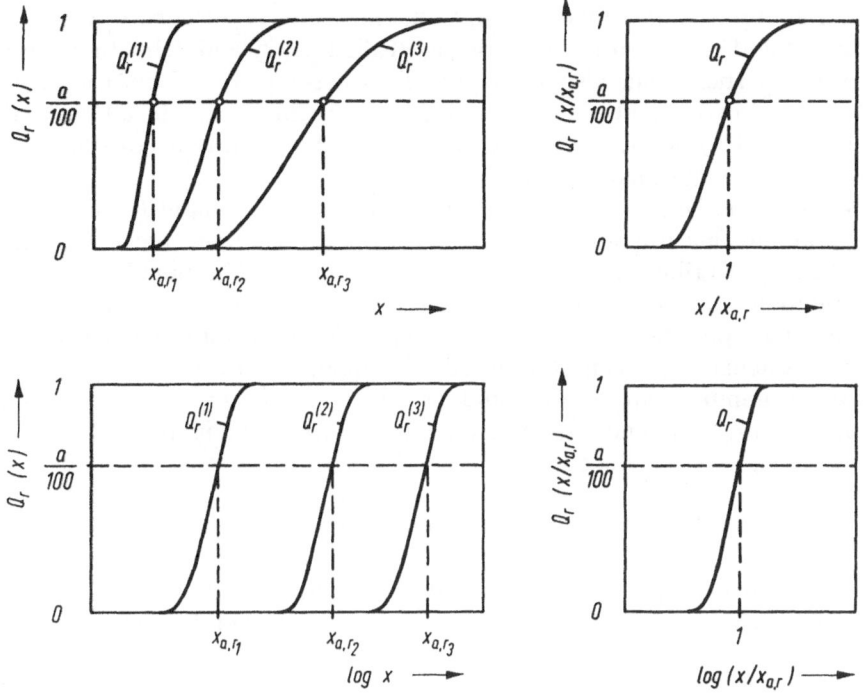

Figure 2.9 Similar distribution functions plotted on linear and logarithmic abscissa scales ($x_{a,r} = x^*$).

We call $x_{(i)}^*$ the characteristic measure of dispersity for the distribution $Q^{(i)}$. The defining equations for similar distribution functions are thus

$$Q^{(1)}(x_{(1)}^*) = Q^{(2)}(x_{(2)}^*) = \ldots = Q^{(i)}(x_{(i)}^*) \qquad (2.25)$$

Similar distribution: $Q^{(i)}(x/x_{(i)})$ identical for all i

If $\log x$ is chosen as the abscissa, similar distribution functions are represented as curves displaced parallel to one another (Fig. 2.9).

2.3 THE SEPARATION OF DISPERSE SYSTEMS

2.3.1 Material balances

Let us consider the separation of a number of particles into two fractions. The separation is to be performed according to the measure of dispersity x. If this is the size of the particle then the separation is called a **classification** and the two fractions are designated as coarse and fine (oversize and undersize). If the separation depends upon the kind of material, then it is called **sorting**; the measure of dispersity x is then usually the density and the fractions are called heavy and light. In a similar way the particles can also be separated according to the electric charge which they carry, and x is now a measure of this. Since in all cases the criteria used to describe the effectiveness of the separation are the same, they can be established using classification as an example.

We shall designate the distribution functions for the input (feed), coarse fraction and fine fraction as $Q_I(x)$, $Q_C(x)$ and $Q_F(x)$ and the corresponding density distributions as $q_I(x)$, $q_C(x)$ and $q_F(x)$. The population of the distribution can be expressed in any way, e.g. by mass or number, depending upon how it has been measured. Therefore the subscript r which designates this can be omitted. The quantities of particles, e.g. their mass or number, are represented as m_I, m_C and m_F. For continuous processes $\dot{m}_I$, $\dot{m}_C$ and $\dot{m}_F$ designate the corresponding flow rates. The material balance has the form

$$m_I = m_C + m_F \qquad \text{or} \qquad \dot{m}_I = \dot{m}_C + \dot{m}_F \tag{2.26}$$

If the relative proportions $c = m_C/m_I = \dot{m}_C/\dot{m}_I$ and $f = m_F/m_I = \dot{m}_F/\dot{m}_I$ are introduced into the material balance equation it then becomes

$$1 = c + f \tag{2.26a}$$

For the particle size grading falling between x_1 and x_2

$$Q_I(x_1, x_2) = cQ_C(x_1, x_2) + fQ_F(x_1, x_2) \tag{2.27}$$

To simplify the presentation we shall omit the arbitrary limits x_1 and x_2 of the size grading; it then follows that

$$Q_I = cQ_C + fQ_F = (1 - f)Q_C + fQ_F = cQ_C + (1 - c)Q_F \tag{2.27a}$$

$$q_I = cq_C + fq_F = (1 - f)q_C + fq_F = cq_C + (1 - c)q_F \tag{2.27b}$$

These equations can be used to find the fractional amounts c and f of the coarse and fine material from the measured distributions according to the relations

$$c = \frac{Q_I - Q_F}{Q_C - Q_F} \qquad f = \frac{Q_I - Q_C}{Q_F - Q_C} \tag{2.27c}$$

In this equation the cumulative distribution function Q can be replaced by the density distribution function q. Whenever direct measurements of the amounts and proportions occurring are available, these equations are very useful for monitoring the measurement of particle distributions.

2.3.2 Grade efficiency and grade efficiency curve, cut size and sharpness of cut

The grade efficiency and the associated grade efficiency curve are used to describe a classification. The grade efficiency $G(x)$ indicates, for a differential x interval, which part of the feed goes into the coarse fraction:

$$G(x) = \frac{cq_C(x)\,dx}{q_1(x)\,dx} = \frac{cq_C(x)}{q_1(x)} \tag{2.28}$$

The grade efficiency is independent of the measure of population used (mass, surface area, number etc.) because it is the ratio of the amounts of the same particles; the ratio of the numbers is then the same as the ratio of the masses. If, as has been done in Fig. 2.10(a), both density plots $cq_C(x)$ and $fq_F(x)$ are drawn within the density distribution $q_1(x)$, then $G(x)$ can be plotted as the ratio of corresponding ordinates since the relation $q_1 = cq_C + fq_F$ holds. The graph of $G(x)$ (Fig. 2.10(b)) is called the grade efficiency curve.

The concept of grade efficiency was introduced into the technology of mineral dressing by Tromp [96]. Therefore it is sometimes called the Tromp partition index, and the grade efficiency curve is called Tromp's curve. Independently of this, the concept of grade efficiency had already been introduced earlier in the technology of gas cleaning where it was known as the fractional dust separation efficiency. Certain x values are assigned to the grade efficiency curve: $G(x_{50.g}) = 0.5$, $G(x_{25.g}) = 0.25$, $G(x_{75.g}) = 0.75$, $G(x_{0.g}) = 0$, $G(x_{100.g}) = 1.0$ and, in general, $G(x_{i.g}) = i/100$. If the separation is absolutely sharp the grade efficiency curve follows a vertical line, i.e. all $x_{i.g}$ are the same. We call the median value or equiprobable size $x_{50.g}$ of the grade efficiency curve the cut size. It denotes that size x from whose grading $q_1(x)\,dx$ half goes into the coarse fraction and half goes into the fine fraction:

$$fq_F(x_{50.g}) = cq_C(x_{50.g}) \tag{2.29}$$

The **analytical cut size** $x_{a.g}$ is of importance in particle size analyses. It is such that the amounts of misplaced or wrongly classified particles, i.e. the amount of material coarser than this value in the fines and the amount finer in the coarse fraction, are equal:

$$f\{1 - Q_F(x_{a.g})\} = f\int_{x_{a.g}}^{\infty} q_F(x)\,dx$$
$$= c\int_0^{x_{a.g}} q_C(x)\,dx = cQ_C(x_{a.g}) \tag{2.30}$$

To distinguish $x_{50.g}$ from $x_{a.g}$ the former is also known as the **preparative cut size**.

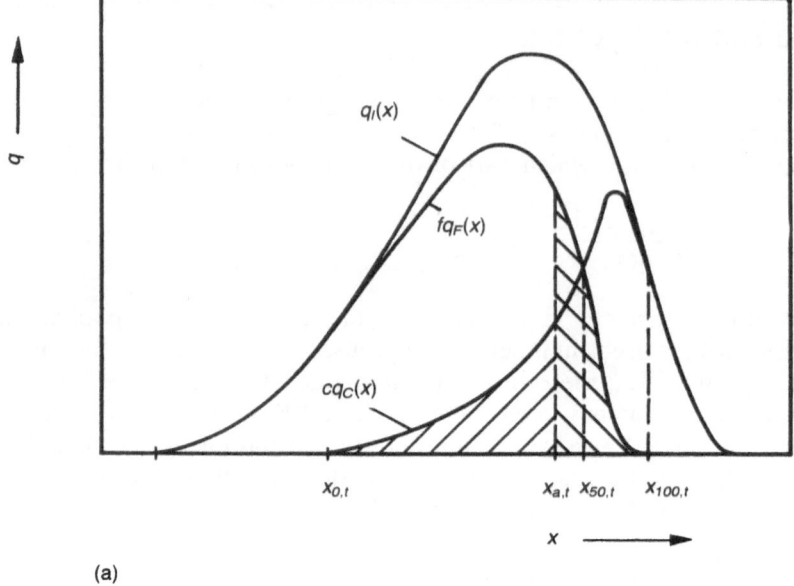

(a)

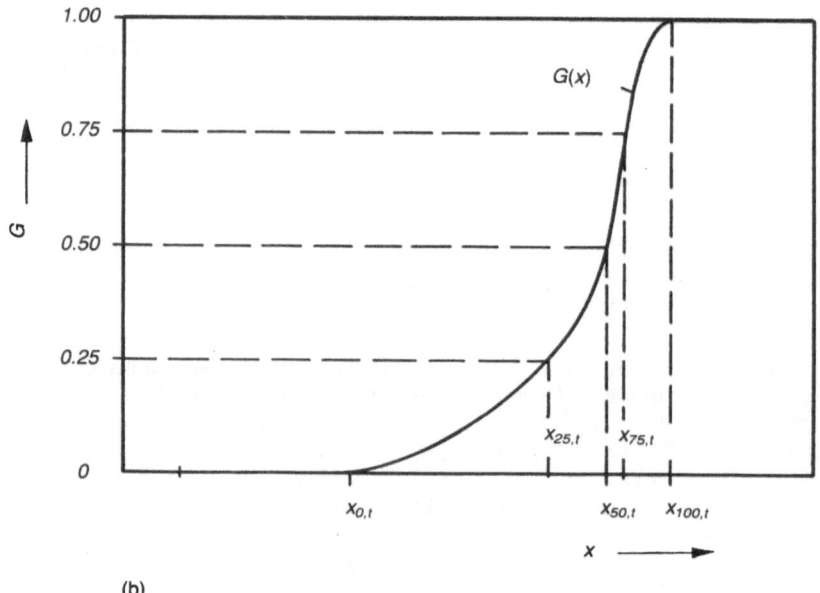

(b)

Figure 2.10 Graphical representation of a separation: (a) density distribution functions $q_I(x)$, $cq_C(x)$ and $fq_F(x)$; (b) grade efficiency curve.

The sharpness of cut, which is an indication of the quality of the separation, can be expressed in terms of a 'sharpness index' $x_{i,g}/x_{j,g}$; the usual ratio is $\kappa = x_{25,g}/x_{75,g}$. Other ways of defining the sharpness index, commonly used in mineral dressing technology, are the *écart probable* or *écart Terra*

$$E_g = \tfrac{1}{2}(x_{75,g} - x_{25,g})$$

Mayer's expression

$$E_M = x_{90,g} - x_{10,g}$$

and the 'imperfection'

$$Im = \frac{x_{75,g} - x_{25,g}}{2x_{50,g}} = \frac{E_g}{x_{50,g}}$$

If grade efficiency curves plotted as a function of $\log x$ run parallel to one another, we refer to them as similar. Such families of curves have the same sharpness indices and the same 'imperfection'. Differences such as E_g and E_M are then only meaningful if separations which occur at approximately the same cut size are being compared with one another, as for example when materials are separated according to their density. If, however, we are comparing sifting performed at $x_{50,g} = 1$ mm with sifting performed at $x_{50,g} = 1\ \mu$m, then a difference of $E_M = 30\ \mu$m signifies an extremely high sharpness of cut in the first case and an extremely low sharpness of cut in the second case.

In practice the κ values achieved are in the following ranges: sharp laboratory separations, $0.8 \leqslant \kappa \leqslant 0.9$; sharp industrial separations, $0.6 \leqslant \kappa \leqslant 0.8$; ordinary industrial separations, $0.3 \leqslant \kappa \leqslant 0.6$. The κ value is a characteristic of only the central range of the grade efficiency curve. Often what happens does not depend so much on this as on the coarse and fine branches of the curve. With abrasives, for instance, a few coarser particles can spoil a size cut because they cause scratches that cannot be tolerated. In circuit grinding the purpose of classification is to prevent recycling of material already sufficiently finely ground and to direct the material which still requires grinding back into the milling zone. If the coarse fraction still contains too many fines then these are unnecessarily ground finer, and they cushion the stressing of the coarser particles and cause an agglomeration that is often very troublesome. Here the result depends less on the κ value than on the grade efficiency curve's approaching the abscissa as closely as possible and not becoming horizontal at some finite grade efficiency such as $G = 0.2$ or $G = 0.1$. Exactly this disadvantage is suffered by many commercial air classifiers and elutriators.

Figure 2.11 shows some examples of grade efficiency curves. Separations b_1, b_2, e_1 and e_2 are sharp.

Separation e_2 is sharp at a cut size of 10 μm but curves upwards again at fine particle sizes. The fine particles under 8 μm agglomerate or stick to the larger particles and thus behave as coarse material. With all continuously operated separations the sharpness of cut is reduced if the feed rate is increased above a certain limit. Beyond a definite loading the separation breaks down completely. In the extreme case the grade efficiency curve becomes horizontal (curve f) and pure partitioning takes place: the same proportion is separated from each fraction (size grading) x. Such a horizontal, i.e. ideal, partition curve is aimed for when splitting a sample.

We distinguish between the separation intended at a particular particle size or other measure of dispersity with as steep a grade efficiency curve as possible when sifting, classifying or sorting, on the one hand, and the separating of solids from fluids (removal of dust, mechanical separation of liquids as in filtering, centrifuging, thickening and clarifying) on the other. In the latter case it is required to remove the whole of the suspended phase from the fluid if possible. The ideal grade efficiency curve is thus horizontal with $G = 1$. In practice all fluid–solid separations have grade efficiency curves which follow a downward course below an x value of $x_{100.g}$. Good cyclones still achieve grade efficiencies (fractional dust separation efficiencies) of $G = 0.9$ at 4 μm (curve d). When comparing the grade efficiency of the cyclone (curve d) with that of the air classifier it should be noted that in the case of the cyclone the loading is smaller by several orders of magnitude (about 10^{-2} to 10 g/kg air compared with about 10^{-1} to 10 kg/kg air). At larger loadings the grade efficiency curve of the cyclone becomes appreciably flatter. Cyclones, especially hydrocyclones, are also used as classifiers.

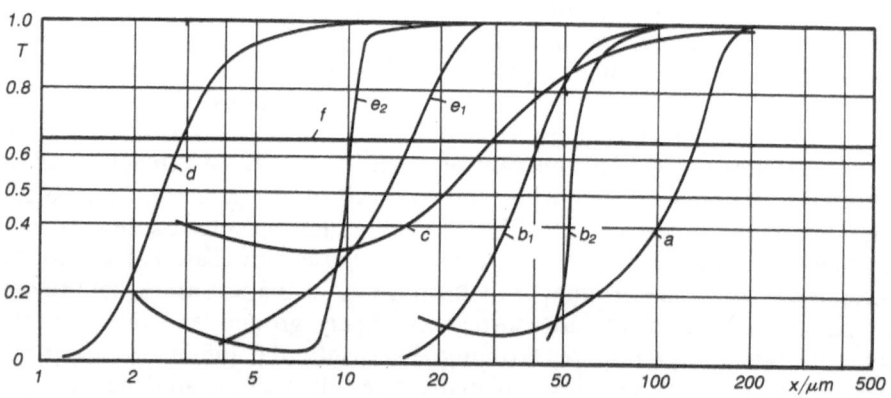

Figure 2.11 Examples of measured grade efficiency curves: curve a, Heyd sifter; curves b, transverse-flow classifier (b_1, heavy loading; b_2, light loading); curve c, Polysius air classifier (3.3 m diameter); curve d, cyclone (Institut für Mechanische Verfahrenstechnik, Karlsruhe; charge 5 mg solids per 1 m^3 crude gas); curves e_1 and e_2, spiral air classifier (Mikroplex MP 400); curve f, sample splitting.

2.4 MIXING OF DISPERSE SYSTEMS

2.4.1 The state of a mixture and the degree of mixing†

A mixture consists of two or more components. The elements of the components are the smallest possible constituents of the mixture. The purpose of a mixing process is to achieve as uniform a dispersion as possible: each sample taken from the mixture should, as far as possible, have the same composition as the whole mixture. Let x be the concentration of a component in a sample and p the corresponding concentration in the whole mixture; if x is then the same as p in each sample the mixture is ideally homogeneous.† The actual concentration p of the mixture is the expected concentration of the sample.

In a real mixture, however, the concentration x in the sample will deviate to some extent from the expected concentration p. For samples of the same size the greater the deviation is on average, the worse is the mixture, whereas the smaller the deviation is, the better is the mixture. The deviations of the sample concentrations from the expected concentration are thus a measure of the quality of the mixture. Since, in characterizing the mixture, it is irrelevant whether the concentration of the sample deviates above or below the expected value, the degree of mixing can be measured by the mean square of the deviations, i.e. the variance of the distribution function for the composition of the sample:

$$\sigma^2(x) = \lim_{m \to \infty} \frac{1}{m} \sum_1^m (x_i - p)^2 \tag{2.31}$$

where m is the number of samples and x_i is the concentration of the ith sample. Since in practice we cannot take an infinite number of samples this value is estimated by the empirical variance of a finite number of random samples:

$$s^2(x) = \frac{1}{m} \sum_1^m (x_i - p)^2 \tag{2.31a}$$

The measures of the degree of mixing frequently quoted in the literature are always combinations of this variance with the variance σ_0^2 of the initial state or the variance σ_F^2 of the possible final state of a mixing process. Some of these measures are given below:

$$\mu_1 = s^2 - \sigma_F^2 \qquad \mu_2 = \frac{\sigma_0^2}{s^2} \qquad \mu_3 = \frac{\sigma_F^2}{s^2}$$

Section 2.4 was written with the cooperation of K. Sommer, Karlsruhe.
†Note that in this section x stands for the concentration of a sample and not for a measure of dispersity.

$$\mu_4 = \frac{\sigma_0{}^2 - s^2}{\sigma_0{}^2 - \sigma_F{}^2} \qquad \mu_5 = 1 - \frac{s^2 - \sigma_F{}^2}{\sigma_0{}^2 - \sigma_F{}^2}$$

$$\mu_6 = 1 - \frac{\log(s^2/\sigma_F{}^2)}{\log(\sigma_0{}^2/\sigma_F{}^2)}$$

For clarity, only the variance itself will be used in the following for designating the degree of mixing.

Let us assume that in the initial state the two components of a mixture are still completely separated from each other. A sample, which should not extend over and beyond the interface, will contain either one component ($x = 1$) or the other ($x = 0$). The variance of this sample is

$$\sigma_0{}^2(x) = p(1 - p) \tag{2.32}$$

This variance is independent of the size of the sample and of the size of the elements of the individual components.

An **ideal homogeneous mixture** gives samples with a constant concentration. Its variance is zero:

$$\sigma_F{}^2(x) = \sigma_{ideal}^2(x) = 0 \tag{2.33}$$

This ideal homogeneity is realized in the regular arrangement of the elements of the mixture's components. **Ideal regular mixtures** do not occur in practice. Even real crystals contain flaws which are systematic random deviations from regularity.

The regular homogeneous state of mixing of real mixtures is the **uniform random mixture** or **stochastic homogeneity**. Its definition is based on the uniform distribution of the probability for the occurrence of an event. It can be defined, with respect to location, for a given space or, with respect to time, for a series of random events. In terms of the spatial definition a uniform random mixture occurs if no local part of the space under consideration is distinguishable from any other such part, i.e. if, for example, the probability of finding one or several elements of the mixture in any particular domain within the space for all identical elements of a component and for all equally large domains within the space in question is equally great. Strictly speaking, this definition is valid only for the spatial distribution of points; for particles occupying space it is approximately true if their concentration by volume is very small. At larger concentrations the particles mutually deprive one another of space. The random mixture then builds itself up successively from particle to particle. A theoretical statistical representation of the process by which a random mixture is formed has not yet been found. If a uniform random mixture is to be produced and maintained a state of motion must be realized in which no selective forces and selective movements occur.

Table 2.4 Variances appertaining to stochastic homogeneity

Measure of concentration	Sample size	Regime of validity	Variance of the stochastic homogeneity	Remarks	Row no.
Number	Number of particles n	n = const	$\sigma_R^2(x) = \dfrac{p(1-p)}{n}$		1
Volume	Volume V_p of all particles	V_p = const, $n \approx$ const $v_y \approx v_x$	$\sigma_R^2(X) = \dfrac{P(1-P)}{V_p}\{Pv_y + (1-P)v_x\}$	After Stange	2
	Total volume $V = V_p$	Suspensions $v_y \ll v_x$ $n \approx$ const	$\sigma_R^2(X) = P(1-P)^2 \dfrac{v_x}{V}$	After Stange: however, contradicts $v_y \approx v_x$	3
	Total volume $V = V_p$	Suspensions $P \ll 1$; V = const	$\sigma_R^2(X) = P\dfrac{v_x}{V}$	After Pawlowski	4
	Volume V_p of all particles	$\dfrac{v_x}{v_y} = \dfrac{i}{j}$; $V_p = \lambda j v_x$ = const	$\sigma_R^2(X) = P(1-P)\dfrac{jv_x}{V_p}$	After Sommer	5
	Volume V_p of all particles	$\dfrac{v_x}{v_y} = i$; $V_p = \lambda v_x$ = const	$\sigma_R^2(X) = P(1-P)\dfrac{v_x}{V_p}$	For large V_p; also valid for $V_p = \lambda v_x$	6
	Total volume V	Suspensions $\dfrac{v_x}{v_y} = i$; therefore:	$\sigma_R^2(X) = P(1-P)\dfrac{v_x}{V}$	For large V; also valid for $V \neq \lambda v_x$	7

Solutions which have been proposed in the literature for the variance of stochastic homogeneity under various sampling conditions or on the basis of specified assumptions about the mixed material are listed in Table 2.4. Here the concentration by number is designated by x and p, the concentration by volume by X and P, and the total volume of all particles in the sample by V_p. For the concentration by number x the variance can be specified exactly for samples containing a constant number of particles n (row 1). For samples of constant V_p, Stange [89] gives an approximate formula for the case where the volumes v_x and v_y of the grains of the components being mixed are approximately the same size (row 2). This formula is also frequently applied to concentrated suspensions where, however, the 'grain volume' of the liquid (a molecule of liquid!) is very much smaller than the grain volume of the granular component, which contradicts Stange's original derivation (row 3). As opposed to this, Pawlowski [74] gives an exact formula for suspensions with an extremely low concentration of solids (row 4). Recent work by Sommer has provided the exact solution for the particular problem when the samples taken contain a constant total volume of particles V_p; here i, j and λ are positive whole numbers (row 5). By introducing the simplification that the grain volume v_x of the first component is an integral multiple of the grain volume v_y of the second component (row 6), a formula for the variance of the stochastic homogeneity of suspensions of any given concentration by volume can readily be derived (row 7). Comparison with the formulae 3 and 4 shows that formula 7 becomes that of Pawlowski for small concentrations $P \ll 1$. It is distinguished from Stange's formula by the factor $(1 - P)$.

2.4.2 The variance of the concentration distribution during the mixing process

The variance σ_0^2 of the initial state, in which the components are completely separated, and the variance σ_R^2 of the stochastically homogeneous final (residual) state are functions of the composition of the sample and the properties of the components only and thus cannot be used on their own for assessing the performance of a mixer; for this purpose only the development of the variance over the period between the initial and the final state of the mixture is suitable.

In order to simplify the matter, we shall limit ourselves to a description of a mixing system in which the local concentration distribution can be adequately represented by a single locating coordinate such as the axial position in mixing tubes. Then the concentration profile $P(r, t_0)$ in the initial state of time $t = t_0$, for instance, is as shown in the upper part of Fig. 2.12. As time passes $(t > t_0)$ the concentration distribution will even itself out. If we repeat this mixing experiment under identical initial conditions then, on average, a characteristic concentration profile $P(r, t)$

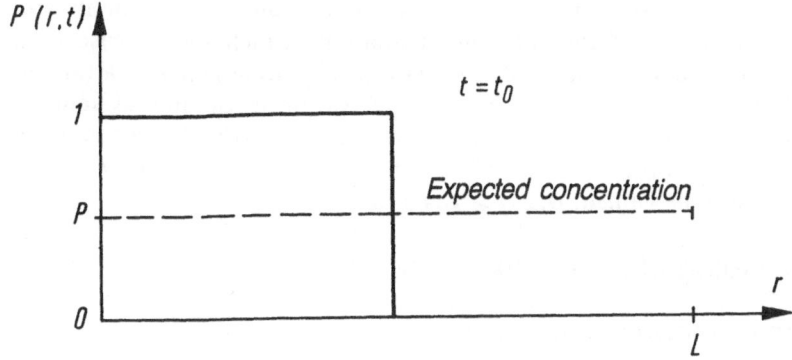

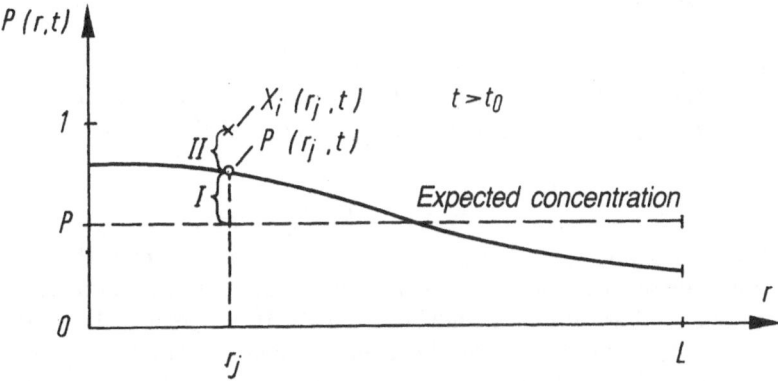

Figure 2.12 Concentration profiles in a mixer.

will emerge at time t. If, however, we consider the concentration $X(r_j, t)$ at point r_j and time t, this will vary about the average value $P(r_j, t)$ at point r_j and time t from experiment to experiment. The variation from the expected value P is thus influenced by two fluctuating variables, i.e. the systematic deviation $P(r_j, t) - P$ and the random variation $X(r_j, t) - P(r_j, t)$. In Fig. 2.12 these fluctuating variables are designated by I and II respectively. If the concentrations are measured, the inaccuracy of the measurement must also be taken into account. The overall variance $\sigma^2(t)$ of the composition of the sample which would result from an infinite number of repetitions of the mixing process and from taking samples at points distributed uniformly over the whole mixing chamber is thus

$$\sigma^2(t) = \sigma_M^2 + \sigma_R^2(t) + \sigma_{\text{syst}}^2(t) \tag{2.34}$$

where $\sigma_M{}^2$ is the variance due to the inaccuracy of the measurements and $\sigma_R{}^2(t)$ is the variance of the random fluctuations which in the case of a random mixer becomes that of the stochastic homogeneity after an infinitely long mixing time. $\sigma_{syst}^2(t)$ is the variance of the systematic concentration distribution and is the appropriate criterion for assessing a mixer. The mixing process is complete when, after a mixing time t, the systematic deviations vanish: $\sigma_{syst}^2(t > t_M) = 0$.

2.4.3 Evaluation of mixing experiments

(a) Mixture of known composition

Let us consider the course of a mixing process. At time $t = 0$ the variance has its maximum value

$$\sigma^2(t = 0) = \sigma_0{}^2 + \sigma_M{}^2 \tag{2.34a}$$

With increasing time a systematic levelling out takes place. After a mixing time t_M the uniform random mixture will be obtained. Then at all locations $P(r_j, t_M) = P$ and $\sigma_{syst}^2(t_M) = 0$. The resulting variance at time t_M is then

$$\sigma^2(t_M) = \sigma_R{}^2(t_M) + \sigma_M{}^2 \tag{2.35}$$

We shall first assume $\sigma_M{}^2 \ll \sigma_R{}^2$; the mixing time t_M can then be measured. Of course, the empirical variance s^2 of a finite number of random samples is determined rather than $\sigma_R{}^2$. If we proceed on the assumption that the composition of the samples taken from the random mixture is normally distributed – which is permissible for sufficiently large samples – then the values of $m(s^2/\sigma_R{}^2)$ found for this mixture are distributed according to the χ^2 distribution, where $\chi^2 = m(s^2/\sigma_R{}^2)$. By consulting the tabulated values of this distribution, it is possible to ascertain the upper and lower confidence limits for $m(s^2/\sigma_R{}^2)$, and hence those $s_u{}^2$ and $s_l{}^2$ for s^2, such that the value of χ^2 will fall within this range with a probability of, say, 95%.

The change in $s^2/\sigma_R{}^2$ with time is plotted in Fig. 2.13. At $t = 0$ the ratio $\sigma_0{}^2/\sigma_R{}^2 = n$, where n is the number of particles in the sample or, depending on the type of statistical argument adopted, the ratio of the total volume of the particles in the sample to that of its elements. At $t = t_M$

$$\frac{\sigma^2(t_M)}{\sigma_R{}^2(t_M)} = 1 + \frac{\sigma_M{}^2}{\sigma_R{}^2(t_M)} \approx 1 \tag{2.36}$$

since it has been assumed that $\sigma_M/\sigma_R \ll 1$. At times $t > t_M$ the measured values of $s^2/\sigma_R{}^2(t_M)$ will fall with 95% probability within the confidence limits

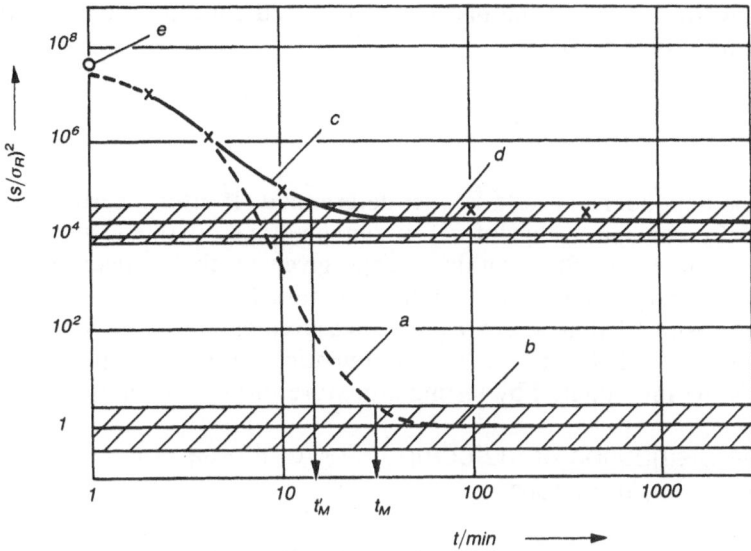

Figure 2.13 Changes in the variance ratio $(s/\sigma_R)^2$ during a mixing process: curve a, presumed course for $(\sigma_M/\sigma_R)^2 \ll 1$; curve b, calculated variance σ_R^2 for the stochastically homogeneous final state; curve c, measured trend for $(\sigma_M/\sigma_R)^2 \gg 1$; curve d, variance σ_M^2 due to the inaccuracy of the measurements; e, theoretically determined initial value $\sigma_0^2 = n\sigma_R^2$.

$$\left\{\frac{s_l}{\sigma_R(t_M)}\right\}^2 < \left\{\frac{s}{\sigma_R(t_M)}\right\}^2 < \left\{\frac{s_u}{\sigma_R(t_M)}\right\}^2$$

The mixing process is then complete.

For $\sigma_M^2 \gg \sigma_R^2$, after a time t_M' the experimental plot of $s^2(t)$ will fall between the confidence limits for σ_M. We can determine only the time t_M' as the mixing time (Fig. 2.13). It is certain, however, that the time t_M necessary to attain a uniform random mixture must at least equal t_M'. How large t_M actually is remains open.

In practice it is frequently necessary to obtain as high a homogeneity as possible, i.e. the uniform random mixture; in this case a sufficiently exact analytical procedure must be found. If this requirement cannot be fulfilled, the indices for the degree of mixing mentioned at the outset are pointless.

(b) Mixtures of unknown composition

If the composition p or P of the mixture is not known the variance σ_R^2 of the composition of the samples from the uniform random mixture cannot be calculated. Despite this, in many cases we may want to know whether we actually have a uniform random mixture. The empirical variance based on the average value of the particular measured concentrations can then be used as an estimator of σ_R^2:

$$s^{*2} = \frac{1}{m-1} \sum_1^m (X_i - \bar{X})^2 \tag{2.37}$$

The time dependence of s^{*2} is then determined. If after a certain time t_M a statistically constant value of s^{*2} is obtained, i.e. if the values of s^{*2} lie with a frequency of 95% within a range given by the χ^2 distribution, then it is assumed that a uniform random mixture has been achieved. Here it must be assumed that at $t > t_M$ the measurement error s_M^2 is very much less than s^{*2} and that there is no systematic demixing. In particular cases this must be investigated by testing a mixture of known composition.

(c) The distribution of the composition of the samples from a uniform random mixture

If samples containing n particles are taken from a uniform random mixture of composition p then the distribution of their composition is the binomial distribution

$$P(n, p, n_x) = \binom{n}{n_x} p^{n_x}(1 - p)^{n - n_x} \tag{2.38}$$

where n_x is the number of particles of component (x) in the sample. If the number of particles n in the sample is large then two approximations to the binomial distribution can be used. The Poisson distribution

$$P(n, p, n_x) = \frac{(np)^{n_x}}{n_x!} \exp(-np) \tag{2.39}$$

is valid provided not only that $n \gg 1$ but also that $p \ll 1$ so that np is always a small number; its variance is $\sigma_x^2 = p(1 - p)/n \approx p/n$. The standardized normal distribution

$$h(t) = \{\sqrt{(2\pi)}\}^{-1} \exp\left(-\frac{t^2}{2}\right) \tag{2.40}$$

with $t = (x - p)/\sigma_x$ and $\sigma_x^2 = p(1 - p)/n$ can be used if both n_x and n_y, as well as n, are very much greater than unity.

(d) Sample size

The sample size n selected must be large enough to ensure that the deviation $\Delta x = |x - p|$ will fall with a given probability within predetermined limits $\pm \Delta x_g$. The corresponding relative deviation about the expected value p is $f_x = \Delta x_g / p$. The following expressions, in which $\Delta x_g / \sigma_x$ has been replaced by ρ, give the required sizes of the samples:

Poisson distribution $\qquad n_{\min} = \dfrac{\rho^2}{f_x^2} \dfrac{1}{p}$ $\qquad\qquad$ (2.41)

normal distribution $\qquad n_{\min} = \dfrac{\rho^2}{f_x^2} \left(\dfrac{1}{p} - 1 \right)$ $\qquad\qquad$ (2.42)

The probability $P(\rho) - P(-\rho)$ that $\Delta x < \Delta x_g$ or $\Delta x / p < f_x$ can be found from tables of the binomial, Poisson and normal distributions.

2.5 PACKING OF GRANULAR SOLIDS

2.5.1 The states of packing systems

The idea of a regular arrangement of particles is often associated with the concept of packing. For example, we speak of a cubic or hexagonal densest packing of spheres. However, the concept of uniform random packing signifies an arrangement of particles which, although stochastically uniform, is not absolutely regular. The loose expression **packed beds** is used regardless of particular arrangements. Here **packing** will be used as a general term for designating the state of all those systems of disperse matter in which the particles mutually fix themselves in position and thereby have only a restricted mobility.

Examples of packings according to this definition are agglomerates, bulk solids, filter cakes, soils, fixed beds and, within certain limits, moving beds and fluidized beds. These systems of disperse matter have certain common properties and modes of behaviour which can be uniformly represented. A packing can remain at rest or be deformed and moved around. If the mobility of the particles increases and the density of the packing decreases the system eventually changes into a state in which their location and movement are determined at least as much by other effects, such as hydrodynamic forces, reactions with a boundary wall or inertial forces, as by mutual interactions. This state is reached with a high probability when the average distance between the centres of the particles is twice as large as the average particle diameter. At this stage the volumetric concentration of the particles is less than 10%.

We thus define a packing as a disperse system in which the mean distance between neighbouring particles (from surface to surface) is generally smaller than the particle diameter. The disperse system of greater

Section 2.5 was written with the co-operation of M. H. Pahl and H. Schubert, both of Karlsruhe.

mobility can be distinguished from a packing by calling it a dilute suspension or a dilute **particle–fluid mixture**. There is a continuous transition between a packing and a dilute particle–fluid mixture. Fluidized beds occupy such a transitional position.

Some states of a particle system are shown schematically in Fig. 2.14. In the fixed bed (Fig. 2.14(a)) the particles are at rest. The void space can be filled by one phase (Fig. 2.14(b_1)) or by two phases (Fig. 2.14(b_2)). The phases can be either still or flowing. If a flow is induced through a fixed bed from the bottom to the top without hindering its expansion it changes into a fluidized bed (Fig. 2.14(c)) at a certain flow rate. A packing can be deformed like a solid body (Fig. 2.14(d)) or it can flow like a liquid (Fig. 2.14(e)). A stream can pass through a moving bed (Fig. 2.14(f)) as it moves and thus wholly or partially fluidize it.

The following measurements or factors serve to characterize a packing: the size or size distribution of the particles, the particle shape parameters,

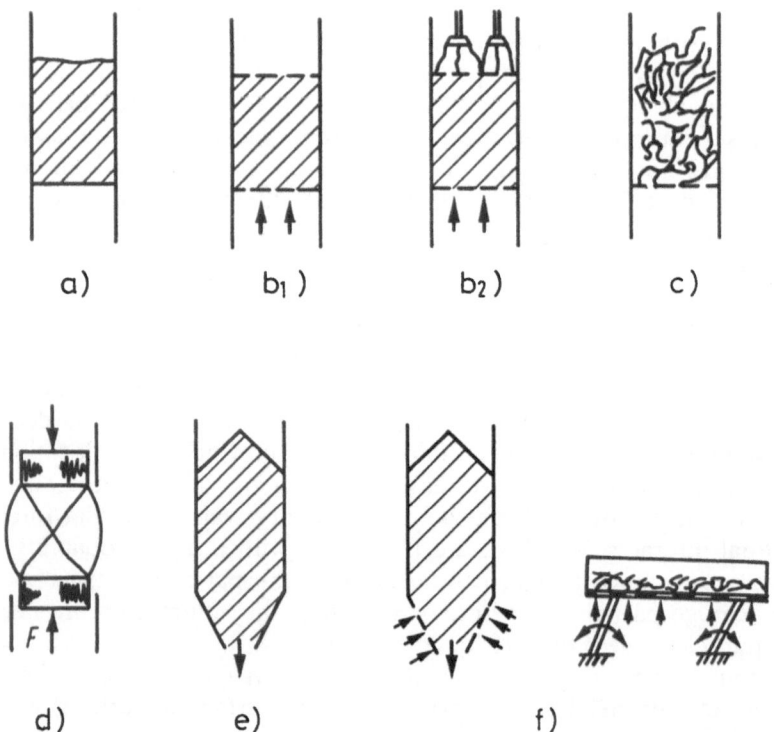

Figure 2.14 States of packing systems: (a) fixed bed; (b) permeated fixed bed; (c) fluidized bed; (d) deformation; (e) flow out of a bunker; (f) permeated moving bed.

the voidage, the structure of the packing, the nature of the liquid filling the voids, the capillary pressure and the forces transmitted within the packing.

2.5.2 The particle and void system

The particles forming a packing are characterized by a location parameter of the particle assembly (e.g. a mean equivalent diameter), by distribution parameters of the assembly (e.g. the standard deviation) and by particle shape factors. The system of voids can be represented in terms of the porosity ε and the pore size distribution or the mean pore size. ε is the ratio of the volume V_ε of voids to the volume V of the packing: $\varepsilon = V_\varepsilon/V$.

Some empirical values of porosities are given in Table 2.5. A pore is generally understood to be a channel running through the packing. If one proceeds from any pore opening at the surface of the packing then there are at least three openings available for the continuation of the channel between each layer of particles, and thus after ten layers there are some $3^{10} = 6 \times 10^4$ possible channels. Which channel is considered to be a pore depends upon the measuring procedure used.

2.5.3 Packing structure

Regular packing structures are defined mainly for packings of spheres having the same diameter. Some packings of this type are listed in Table 2.6, in which k is the coordination number, i.e. the number of contact points which each particle has with its neighbours. Such regular packings can be produced with spheres of exactly the same diameter. Even when there are only slight variations in diameter random packing is found a few layers after the initial layer [97] (see Fig. 2.15).

The essential feature of uniform random packing is that the probability of finding the centroid of a particle at any given location within the space

Table 2.5 Empirical values of porosities

Poured materials	0.3–0.6
After shaking down	0.25–0.45
Raschig rings	0.55–0.75
Broken mica	0.8–0.9
Fluidized bed	0.4–0.9
During pneumatic conveying	0.95–(almost) 1.0
Filter cakes	0.5–0.9
Filter cakes (filter press)	0.4–0.9
Ceramic filters (porous stone)	0.2–0.3
Sintered glass filters	0.09–0.15
Briquettes	0.05–0.20
Pellets	0.25–0.40

Table 2.6 Coordination number k and porosity ε of regular packings of spheres

Designation	k	$1 - \varepsilon$	ε
Primitive cubic	6	$\pi/6$	0.477
Orthorhombic	8	$(\pi/6)(2/\sqrt{3})$	0.395
Tetragonal spheroid	10	$(\pi/6)(2/\sqrt{3})^2$	0.302
Rhombohedral (hexagonal)	12	$(\pi/6)(2/\sqrt{2})$	0.259
Tetragonal	12	$(\pi/6)(2/\sqrt{2})$	0.259

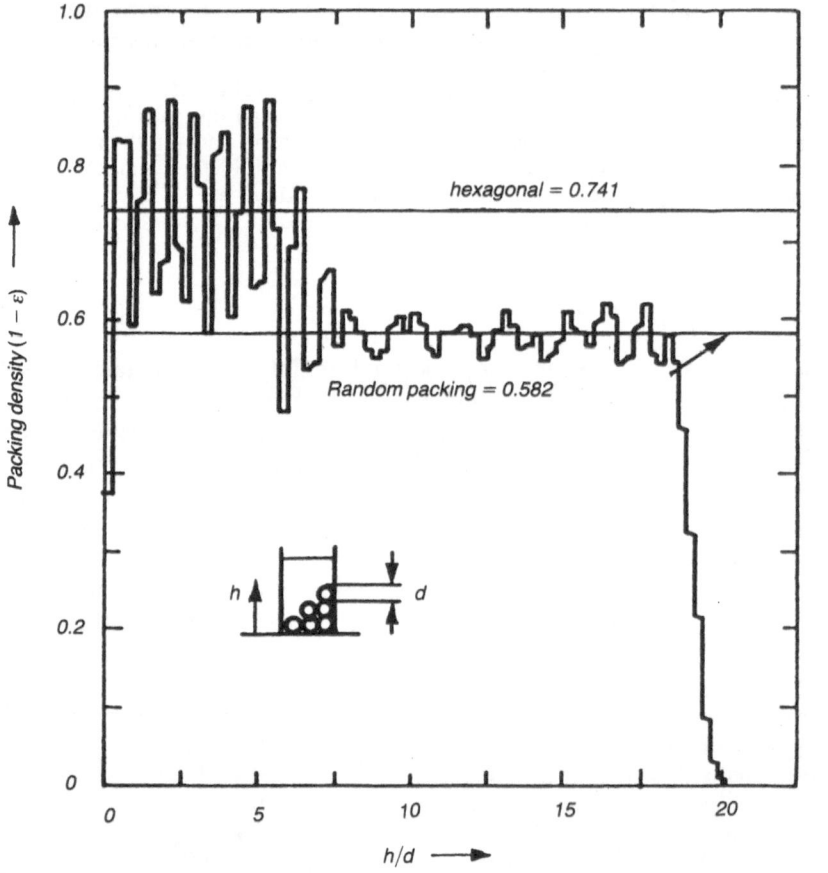

Figure 2.15 The influence of a boundary wall on the packing of spheres [97].

of the packing is equally great for all similar particles and all locations. The spatially random arrangement which thus ensues is assumed by points which occupy no space of their own. The law governing the formation of a uniform random arrangement for particles occupying space is expressed by stating that no spatial domain is distinguishable from any other and that a particle is equally likely to be found in any possible position. Likewise there is an equal probability as regards the orientation of the particle within the space (cf. the definition of a random mixture on p. 36). If a random packing is made up of particles of differing sizes or from two different particle assemblies, then the relevant probabilities for the various particles are determined by the frequency with which they occur in the whole packing.

If we take samples of a given volume from the random packing then their porosity ε_{sa} will vary randomly from sample to sample. Its expected value is the porosity of the whole packing:

$$E(\varepsilon_{sa}) = \varepsilon \qquad (2.43)$$

The variance of the porosity ε_{sa} of the samples can be calculated using Sommer's statistics of mixing. If the packing consists of particles of equal size of volume v_p, then it follows that

$$\sigma_{\varepsilon_{sa}}{}^2 = \frac{\varepsilon(1 - \varepsilon)v_p}{V_{sa}} \qquad (2.44)$$

where V_{sa} is the volume of the sample. Sommer's mixing statistics are also applicable to packings of non-uniform particles.

To examine a packing we can make use of sectional areas taken through the undisturbed packing as shown in Fig. 2.16. In any given cross-section the discrete sectional areas of the particles are embedded in the continuous

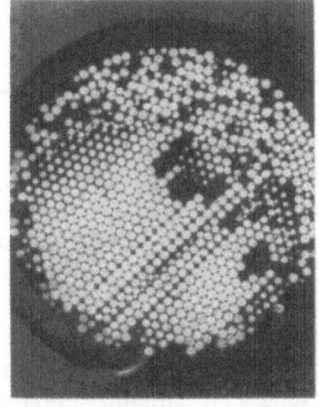 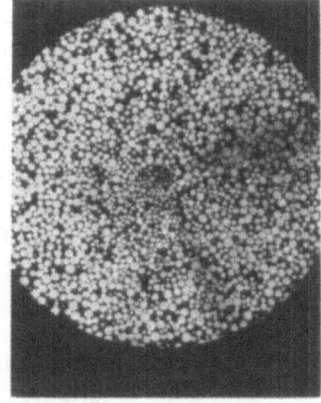

(a)   (b)

Figure 2.16 Sections through packings of uniformly sized lead spheres: (a) partly systematic arrangement of particles; (b) random arrangement of particles.

cross-sectional area of the pore space. If ε_A is the ratio of this area to that of the total cross-section then it follows that for a uniform random mixture the expected value $E(\varepsilon_A)$ of the porosity ε_A per area is

$$E(\varepsilon_A) = \text{const} = \varepsilon \tag{2.45}$$

This is a necessary but not sufficient condition for testing a packing with regard to its uniform random structure. For many applications (e.g. in mineralogy) the relation between the distribution by number of the sectional areas and that of the particle sizes is of interest. If Θ is the diameter of the circular area of sliced spherical particles of diameter x, $z(\Theta)$ is the frequency distribution of these sectional diameters and $q_0(x)$ is the frequency distribution, and the first moment of the particles involved is $M_{1,0}$, then, according to Wicksell [102], the relation between the density distributions is given by

$$z(\Theta) = \frac{\Theta}{M_{1,0}} \int_{\Theta}^{x_{max}} \frac{q_0(x)}{(x^2 - \Theta^2)^{1/2}}\, dx \tag{2.46}$$

If the packing consists of particles of equal size, then the relation is reduced to

$$z(\Theta) = \frac{\Theta}{x(x^2 - \Theta^2)^{1/2}} \tag{2.47}$$

The moments of the $q_0(x)$ distribution can be calculated from the moments of this distribution [12]. Recent considerations of this relation have cast some doubt on Wicksell's theory.

2.5.4 Liquid filling of voids

The voids of a packing made up of solid particles may be filled partly with air and partly with liquid. The liquid may or may not wet the solid. Such systems can be dealt with in a uniform way if it is possible to distinguish between the wetting and the non-wetting phase. Thus in the system air–water–limestone, for example, we consider water to be the wetting phase and air to be the non-wetting phase, and in the system mercury–air–solid, mercury is the non-wetting phase and air is the wetting phase.

We shall designate the degree of saturation, i.e. the proportion by volume of the wetting phase in the pore space, by S. The ratio of the volume of the liquid to that of the solid is also frequently used to indicate the **residual moisture content** found in the packing; this ratio is designated by the symbol ϕ. Therefore it follows that $\phi(1 - \varepsilon) = S\varepsilon$ is the proportion of the volume of the liquid in the whole volume of the packing. We distinguish between liquid filling capillaries, forming bridges, adhering to surfaces and held internally as follows (Fig. 2.17): **capillary liquid** ϕ_c, **bridging liquid** ϕ_b, **adhering liquid** ϕ_a and **internal liquid** ϕ_i.

The internal liquid is found in the pores of porous solid particles and the adhering liquid clings to the surface of these particles. With wettable solids

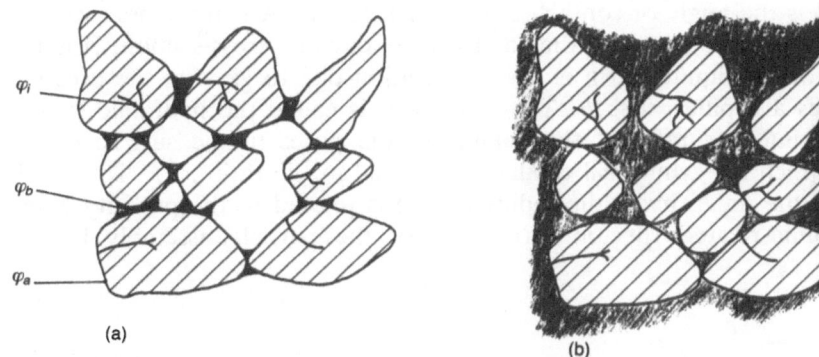

Figure 2.17 Schematic arrangement of the liquid contents of a packing: (a) bridging domain; (b) capillary domain.

in air of ordinary moisture content the adhering liquid is always present. Nevertheless the degree of saturation S is usually still almost zero. The bridging liquid collects itself in the form of bridges where the solid particles touch one another. When these bridges do not coalesce this is known as the 'pendular state'. The upper limit of the degree of saturation for the pendular state depends mainly on ε; at $\varepsilon \approx 0.4$ it lies between 0.2 and 0.3 for packings of uniform grains. If the packing is extensively filled with liquid and the liquid bridges no longer exist then we have the 'capillary state'. For packings of uniform grains and porosity $\varepsilon = 0.4$, this state arises at $S = 0.8$. In the transitional domain or the 'funicular state' liquid bridges and completely filled regions exist alongside one another.

2.5.5 Capillary pressure and porosimetry

The capillary pressure is given by the Laplace equation as the pressure difference between the concave and convex sides of the curved interface between phases:

$$p_c = \gamma\left(\frac{1}{R_1} + \frac{1}{R_2}\right) \tag{2.48}$$

Here R_1 and R_2 stand for the principal radii of curvature of the interface and γ is the surface tension at the interface between liquid and gas. If this equation is applied to the case of the cylindrical capillary then the expression for the capillary pressure becomes

$$p_c = \rho g h = \gamma \frac{P}{A}\cos\delta = \gamma \frac{2}{R}\cos\delta \tag{2.48a}$$

where δ is the angle of contact, P is the perimeter, A is the cross-sectional area of the capillary, R is the radius of its circular cross-section, ρ is the density of the liquid and g is the acceleration due to gravity; the pressure p_c causes a capillary rise h. The ratio A/P is called the hydraulic radius. For packings the ratio of the volume of the pores to the surface of the pores, which for cylindrical capillaries is equal to the ratio A/P, is used. It is called the mean hydraulic radius r_h and is related to the porosity ε and the volume-related specific surface S_v of the particle assembly by the relation

$$r_h = \frac{\varepsilon}{S_v(1 - \varepsilon)} \tag{2.49}$$

The capillary pressure in packings is a function of the degree of saturation. In order to ascertain the path of this function a packing which has been first evacuated and then completely filled with liquid is placed on a diaphragm whose pores are essentially finer than the smallest pores of the packing, the pressure on the system is raised and the amount of liquid expelled is measured at the same time as the pressure. Figure 2.18 shows such a capillary pressure curve. The pressure difference across the packing is set equal to p_c. At $p_c = 0$ it follows by definition that $S = 1$. The pressure must first be raised substantially in order to initiate a desorption. From a certain pressure level onwards the curve becomes flatter but again rises steeply before reaching the final state S_{fin}; this lower limit S_{fin} cannot be overrun because of the remaining liquid that is either isolated or bound to the particles. Curve D_0 describes the desorbing process. If the pressure is lowered again the liquid forces its way back into the packing, which gives rise to the reabsorption curve R; at $p_c = 0$ the value S_a is attained, which is below unity. A renewed desorption produces the curve D and subsequent cycles follow the hysteresis loop R–D–D_0.

A characteristic value is the 'entry suction' p_e, which is given by the point of intersection of the extrapolated steep and flat portions of the desorption curve D_0; the degree of saturation S_e corresponds to this. The value of p_c depends on the surface tension γ, the contact angle δ, the porosity ε, the location parameter $\bar{x}$ of the particle assembly, the distribution parameters ξ_i, and the shape factors $\psi_{j,k}$. In the case of a uniform random packing, dimensional analysis yields the relation

$$\frac{p_e \bar{x}}{\gamma} = f(\delta, \varepsilon, \xi_i, \psi_{j,k}) \tag{2.50}$$

If we compare only packings with similar particle shapes and size distributions then $\psi_{j,k}$ and ξ_i are constant and no longer appear as variables. They then influence the function f by, for example, affecting the

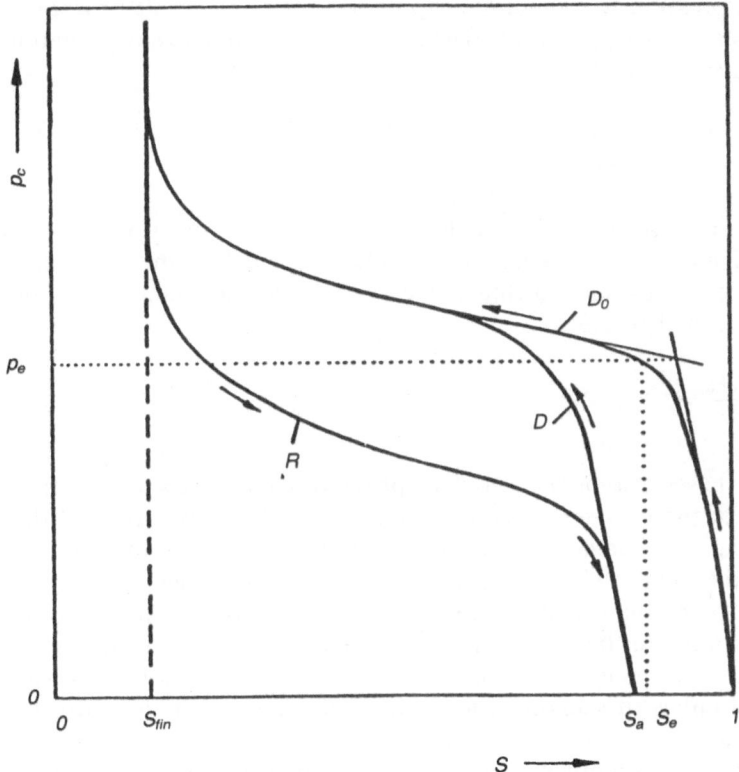

Figure 2.18 Complete capillary pressure diagram for a packing.

factors or exponents which occur in it. The desired function then has the form

$$\frac{p_e \bar{x}}{\gamma} = f(\delta, \, \varepsilon) \tag{2.50a}$$

Measurements by Schubert [24] for packings consisting of spheres of size x with a porosity range $0.33 < \varepsilon < 0.57$ and a contact angle $\delta = 0$ showed that

$$\frac{p_e x}{\gamma} = \frac{6.97(1 - \varepsilon)}{\varepsilon^{0.85}} \tag{2.51}$$

For particle distributions it is recommended, on the basis of ex-perimental measurements, that the Sauter diameter $\bar{x}_2 = M_{1,2}$ is used for x in eqn (2.51). If the mean hydraulic radius $r_h = \varepsilon/\{S_v(1 - \varepsilon)\}$ is substi-tuted for $\bar{x}$ in eqn (2.50a), it becomes $p_e r_h/\gamma = f^*(\delta, \, \varepsilon)$. The application of this expression to cylindrical capillaries gives the result $f^*(\delta, \varepsilon) = \cos \delta$. If this result is applied to packings the Newitt–Kozeny equation [70] is obtained:

$$\frac{p_e r_h}{\gamma} = \frac{p_e \varepsilon}{\gamma S_v (1 - \varepsilon)} = \cos \delta \tag{2.52}$$

The influence of the angle of contact has not yet been sufficiently investigated. For packings of irregularly shaped grains the shape effect is allowed for by introducing a factor A which is found experimentally to range from 1 to 1.3:

$$\frac{p_e r_h}{\gamma} = A \cos \delta \tag{2.52a}$$

The hysteresis of the capillary pressure curve is caused by the changing shape of the cross-sections in the packing and the hysteresis of the angle of contact. The advancing angle of contact is generally larger than the retreating angle. If a pore is enlarged in its middle section and has a larger radius there than at its extremes, then it will be completely emptied at a given pressure but only the lower part will fill up when the applied pressure is reduced because a lower capillary pressure prevails in the central enlarged section. The external pressure must therefore be reduced even more in order to enable the liquid to rise into this section.

By means of mercury porosimetry information can be obtained about the pore size distribution. With this procedure, mercury, as a non-wetting phase, is forced under pressure into an evacuated packing and the amount forced in is plotted against the pressure. A curve is obtained that corresponds to the dehumidifying curve and the course of this curve is interpreted as a measure of the pore size distribution; this is done by ascribing to the capillary or applied pressure a pore size that is given by the formula for the capillary pressure of cylindrical pores. If, therefore, S is chosen as the ordinate and p_c as the abscissa, the capillary dehumidifying curve then represents the distribution function by volume of the pore size attributed to p_c.

The allocation of a pore size to a given value of p_c is problematic, however. First it must be remembered that small values of S correspond to the bridging domain and, with increasing p_c, only the 'vacuum' bridges between the particles become smaller, i.e. it is no longer possible to ascribe p_c to 'pores'. Furthermore, the fact that pore constrictions play a decisive role in mercury porosimetry, as has been demonstrated by Schubert [24] for a plane model (Fig. 2.19), should not be overlooked.

It is assumed that there are spaces of various magnitudes between the circular particles. The sizes of these spaces are designated by the numbers

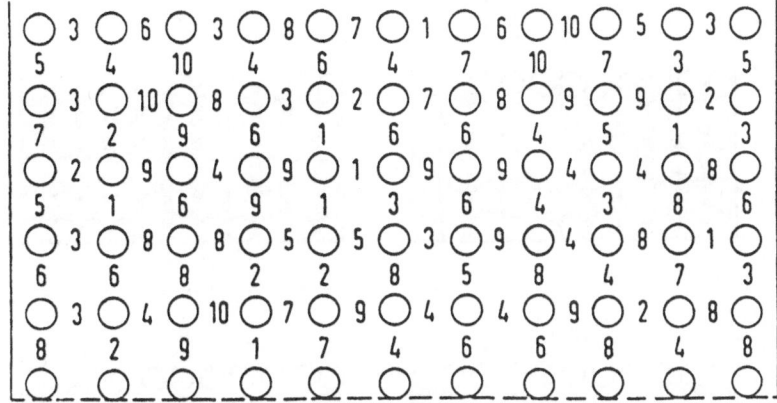

Figure 2.19 Plane model of a packing showing the desorption process [85].

1–10, where 1 denotes the smallest and 10 the largest gap. The numbers are distributed on a random basis. The space between the particles is assumed to be filled with liquid which is to be expelled. In each case, that capillary pressure which can arise in a constriction has to be overcome; the wider the constriction is, the lower is the pressure. Initially, a pressure is applied which just overcomes the capillary pressure of constriction 10, thus exposing the pore lying behind this constriction. From there another constriction 10 leads to a neighbouring pore but further penetration is not possible. If we increase the driving pressure up to the capillary pressure of constriction 9, then a neighbouring pore is opened up. Figure 2.20 shows the result for further increases in the driving pressure (the exposed pores are white). At driving pressures corresponding to constrictions 8, 7, 6, 5 and 4 more channels are opened up. This investigation has two important results which have also been confirmed by experiments on three-dimensional packings.

First, even in the early stages of the desorption, channels which go right through the packing are opened up. This is found to an even greater extent when three-dimensional rather than plane systems are investigated. Thus the liquid is not expelled layer by layer but from these penetrating channels.

Secondly, the displacement process is not governed by the size of the pores but by the size of the constrictions which surround the pores. It is quite possible that large pores are surrounded by small constrictions and therefore are not opened up. Thus porosimetry yields a pore distribution based on the size of the constrictions surrounding the pores. A corresponding conclusion holds with regard to the displacement of the liquid ascribed to the capillary pressure for a given case.

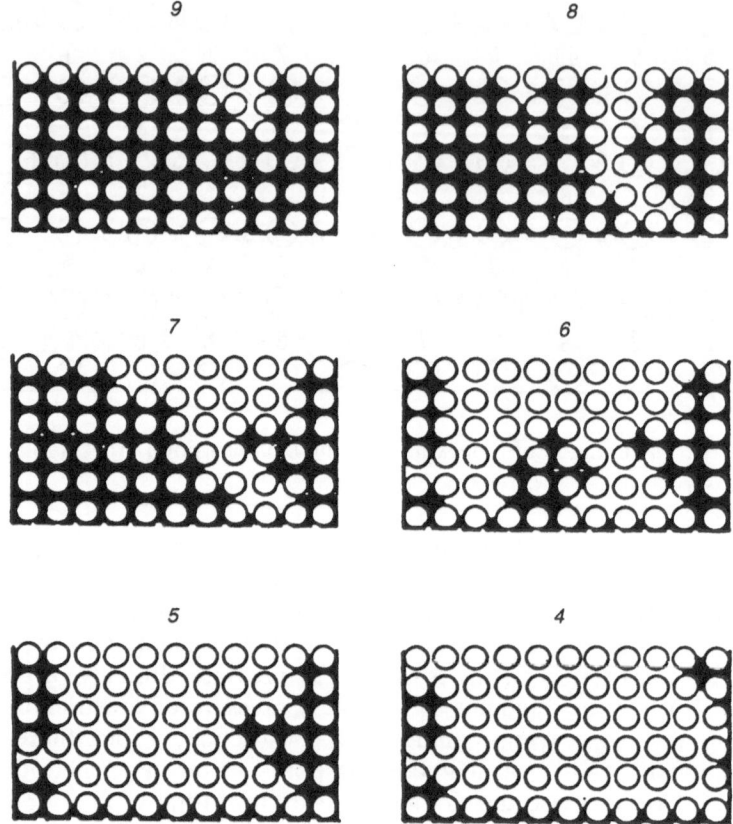

Figure 2.20 Progressive dehumidification of a plane model packing [85].

2.5.6 Forces transmitted in the packing

In many practical circumstances the behaviour of the packing is critically influenced by the forces that can be transmitted within it. These are governed by the stress and the state of motion. The forces must be transferred from particle to particle. In this respect the mutual interactions between the particles, with due regard to the influence of the surrounding phases and the electric field, are the basis for the transmission of forces within the packing.

If the packing is considered as a continuum, then the transmittable forces correspond to normal and shear stresses in sectional planes and can be represented at each point of the packing by the stress tensor. These stresses can then be measured and their characteristic maxima can be used to specify the packing strength. The behaviour of the packing is characterized by the dependence of the stresses on its deformation and rate of deformation. When using such a phenomenological procedure we have to determine the characteristic values for the material of every packing, such as the modulus of elasticity, the velocity of sound, the tensile strength, the shear strength and the rheological properties that affect the transmission of forces when the packing is in a state of motion. Over and above all this, we should attempt to deduce these global properties of the packing from the properties of the particles that form it and from the packing structure and the composition of the phases in the pores.

With regard to the state of motion of the packing we can distinguish three regimes:

1. The packing as a **compact body:** the particles remain to a large extent in mutually fixed positions, such as in an agglomerate for example. The stresses are a consequence of elastic and small plastic deformations such that the particles remain to a large extent bound in the matrix. The behaviour of the material is, above all, a function of the tensile strength and of the shear strength as it starts to yield. Under certain assumed conditions the tensile strength can be estimated on the basis of the properties of the particles, the porosity and the nature of the liquid occupying the voids [85].

2. The packing as a **plastic body:** the packing deforms plastically, i.e. it flows with a small rate of deformation. In this case a rearrangement of the particles takes place and the state of the packing becomes that of steady state flow. With overconsolidated packings the rearrangement is accompanied by a loosening of the structure, and with underconsolidated packings by compaction of the structure. If the packing material is subjected to an imposed deformation the stress tensor can be measured and the behaviour of the packing characterized by particular values. This topic belongs to the mechanics of flowing bulk solids and is treated in section 3.4. The stresses do not depend on the rate of deformation. According to observations so far available this independence persists with coarse materials up to relatively high rates of 1 m/s.

3. The packing as a **flowing medium:** at a higher rate of deformation the stresses start to depend on this rate. The packing then behaves like a fluid. Frictional and inertial forces participate in the transmission of stress. We still know relatively little about this state of a flowing bulk solid.

3

Fundamental physical processes and particle metrology

3.1 THE MOTION OF PARTICLES IN A FLOWING MEDIUM

3.1.1 The forces on a single particle

(a) Synopsis

Let us consider a single particle (Fig. 3.1) suspended in a fluid at a given point in time. Its instantaneous position is determined by its orientation and the position of its centre of gravity. We shall designate its translational velocity by w and its rotational velocity by ω. In the absence of the particle the fluid at the point of time in question has a definite velocity profile with a velocity v at the position of the centre of gravity of the particle. The relative velocity is

$$v_{\text{rel}} = v - w \tag{3.1}$$

If the particle is small with respect to the spatial changes in the field of flow, then the flow can be considered to have a flat velocity profile relative to the particle. In this case v_{rel} denotes the **velocity of approach**, i.e. the uniform velocity with which the fluid approaches the particle (strictly speaking, from an infinite distance).

The accelerations $\dot{v}$, $\dot{w}$ and $\dot{v}_{\text{rel}}$ which must satisfy the equation

$$\dot{v}_{\text{rel}} = \dot{v} - \dot{w} \tag{3.2}$$

can take any direction other than those of the velocities v, w and v_{rel}.

The following types of force can act on the particle.

Section 3.1 was written with the cooperation of Dr J. Raasch.

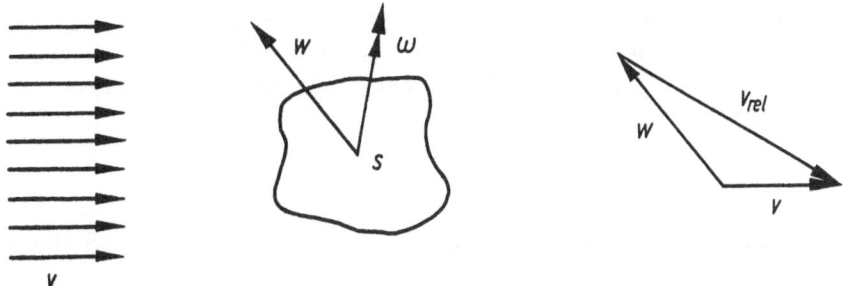

Figure 3.1 Sketch illustrating the definitions of particle velocity w, fluid velocity v and relative velocity v_{rel}.

1. **Field forces:** the best known and most important example is the gravitational force

    $$G = V\rho_p g \tag{3.3}$$

    where V and ρ_p are respectively the volume and the density of the particle and g is the acceleration due to gravity. Electric and magnetic fields of force can also act on the particle.

2. **Aerodynamic and hydrodynamic forces:** as a consequence of the motion of the fluid relative to the particle, it is acted upon, in the most general case, by a torque M and a force F. The latter can be resolved into a component in the direction of v_{rel}, which is the drag force D, and a component perpendicular to v_{rel}, which is the dynamic thrust T_d. The torque M and the component forces D and T_d are not constant quantities. They vary according to whether or not the flow is steady state and depend, *inter alia*, on the Reynolds number and the degree of turbulence in the fluid, on its compressibility and mean free path, on the proximity of boundary walls and other particles, and on the surface roughness and the shape of the particles. These matters will be examined more thoroughly in later sections.

3. **Pressure forces:** in addition to the dynamic forces, and also when there is no relative movement between fluid and particle, pressure forces B can be exerted on the particle. This is always the case whenever a pressure gradient $\operatorname{grad} p$ exists in the field of flow. The general relation is

    $$B = -V \operatorname{grad} p \tag{3.4}$$

    The pressure gradient derived from the Navier–Stokes equations is

    $$\operatorname{grad} p = \rho_f g - \rho_f \dot{v} + \eta \nabla^2 v \tag{3.5}$$

    if it can be assumed that the fluid is incompressible and that no other particles are nearby (suspension with a vanishingly small solids concentration). ρ_f and η stand for the density and viscosity of the fluid respectively. In a quiescent fluid and under the effect of gravity only it follows that

$$\text{grad } p = \rho_f g \tag{3.6}$$

The pressure force in the gravitational field is also called the static thrust (buoyancy).

4. **Inertial forces:** in accordance with d'Alembert's principle an inertial force

$$I = -V\rho_p \dot{w} \tag{3.7}$$

is introduced where V, ρ_p and $\dot{w}$ are respectively the volume, density and acceleration of the particle. If the velocity w of the particle is related to a rotating system of reference, as is the case, for example, when determining the motion of a particle in a centrifuge, two additional inertial forces have to be introduced, i.e. the centrifugal force

$$Z = V\rho_p R\Omega^2 \tag{3.8}$$

and the Coriolis force

$$C = V\rho_p \times 2[w \times \Omega] \tag{3.9}$$

where R is the radius vector of the rotating system of reference and Ω is its angular velocity; $[w \times \Omega]$ is the vector product.

5. **Diffusion forces:** arise either through the direct molecular bombardment of the particle's surface (Brownian movement, radiation pressure) or through equalizing flows which occur in gases as a consequence of a temperature or concentration gradient.

6. **Contact forces:** forces can be transferred by bodily contact between particles and solid walls or between particles alone. In this case it is necessary to distinguish between impact forces, frictional forces and adhesional forces.

(b) The drag force on a sphere in a fluid moving with a constant and uniform velocity of approach

There is as yet no completely general solution to the problem of determining the dynamic forces acting on a particle in a fluid because so many variables are involved. Only a few very special partial problems have been amenable to a theoretical treatment and there the experimental data are inadequate. Therefore, in the first place, far-reaching simplifying assumptions have to be made which will be considered in later sections. These simplifications are as follows:

1. the particle is spherical
2. the particle is rigid
3. the particle has an absolutely smooth surface
4. the particle is at rest
5. solid walls and free surfaces are so far away that in practice they have no influence on the flow
6. the flow path of the approaching fluid is straight
7. the flow is steady state
8. the flow is laminar
9. the flow has a flat velocity profile
10. the fluid is Newtonian
11. the fluid is incompressible
12. the mean free path of the molecules of the fluid is several orders of magnitude smaller than the size of the particles

Under these assumptions the drag force D is a function of the velocity of approach v_{rel}, the viscosity η, the density ρ_f and the particle diameter x_p only. Dimensional analysis shows that two independent dimensionless groups can be formed from these five variables. The first is the Newton number Ne:

$$Ne = \frac{D}{(\pi/4)x_p^2 \times \rho_f v_{rel}^2/2} \tag{3.10}$$

where $(\pi/4)x_p^2$ is the projected area of the particle in the direction of flow and $\rho_f v_{rel}^2/2$ is the impact pressure. The second is the Reynolds number Re which is related to the particle diameter:

$$Re = \frac{x_p v_{rel} \rho_f}{\eta} \tag{3.11}$$

When the fluid approaching a spherical particle is flowing steadily with a flat velocity profile, the law of resistance for the particle then takes the form

$$Ne = C_D(Re) \tag{3.12}$$

or, solving for the drag force D,

$$D = \frac{\pi}{8} x_p^2 \rho_f v_{rel}^2 C_D(Re) \tag{3.13}$$

The function $C_D(\mathrm{Re})$ is known as the drag coefficient. The whole course of the resistance law is reproduced in Fig. 3.2 [22]; it has five ranges.

(i) Stokes range: Re < 0.25

In the range of 'creeping', i.e. viscous, flow the inertial forces in the basic hydrodynamic Navier–Stokes equations can be neglected. In 1851 Stokes solved the problem of viscous flow around a sphere theoretically [91]. The law named Stokes law after him is

$$D = 3\pi\eta x_p v_{\mathrm{rel}} \tag{3.14}$$

or, in the form of eqn (3.12) for the Newton number,

$$\mathrm{Ne} = 24/\mathrm{Re} \tag{3.15}$$

When this is plotted on log–log paper as in Fig. 3.2 a straight line with a gradient of 45° is obtained. Measured values show very good agreement with Stokes law in the range $\mathrm{Re} < 0.25$. The deviations become greater as the Reynolds number increases. If some inaccuracy can be tolerated, this law can still be used up to $\mathrm{Re} = 1$.

(ii) Intermediate range: 0.25 < Re < 10³

In the range above $\mathrm{Re} = 0.25$ there is a rapid increase in the influence of the inertial forces. Eventually the flow no longer closely follows the contour of the sphere but, from about $\mathrm{Re} = 24$ onwards, begins to become detached in the form of individual eddies. At $\mathrm{Re} = 10^3$ this region of separation is developed to its full extent [92]. The course of the drag coefficient plot in the intermediate range has been established experimentally. Numerical solutions of the complete Navier–Stokes equations have been achieved up to a Reynolds number of about 100 showing good agreement with the measured values in this range [42, 49, 52].

(iii) Newton range: 10³ < Re < Re_c

The range above $\mathrm{Re} = 10^3$ is characterized by the fact that up to a critical Reynolds number Re_c the size of the region where the eddies detach themselves behind the sphere barely changes. The physical cause of the separation of the flow itself is that the flow in the boundary layer on the upstream side of the sphere accelerates with a concomitant decrease in pressure energy, while conversely it decelerates on the downstream side of the sphere with an accompanying increase in the pressure energy. However, this recovery is only partial because of the loss of energy due to friction. As a consequence, the flow in the boundary layer is reversed, causing it to separate with the formation of eddies. The point of separation

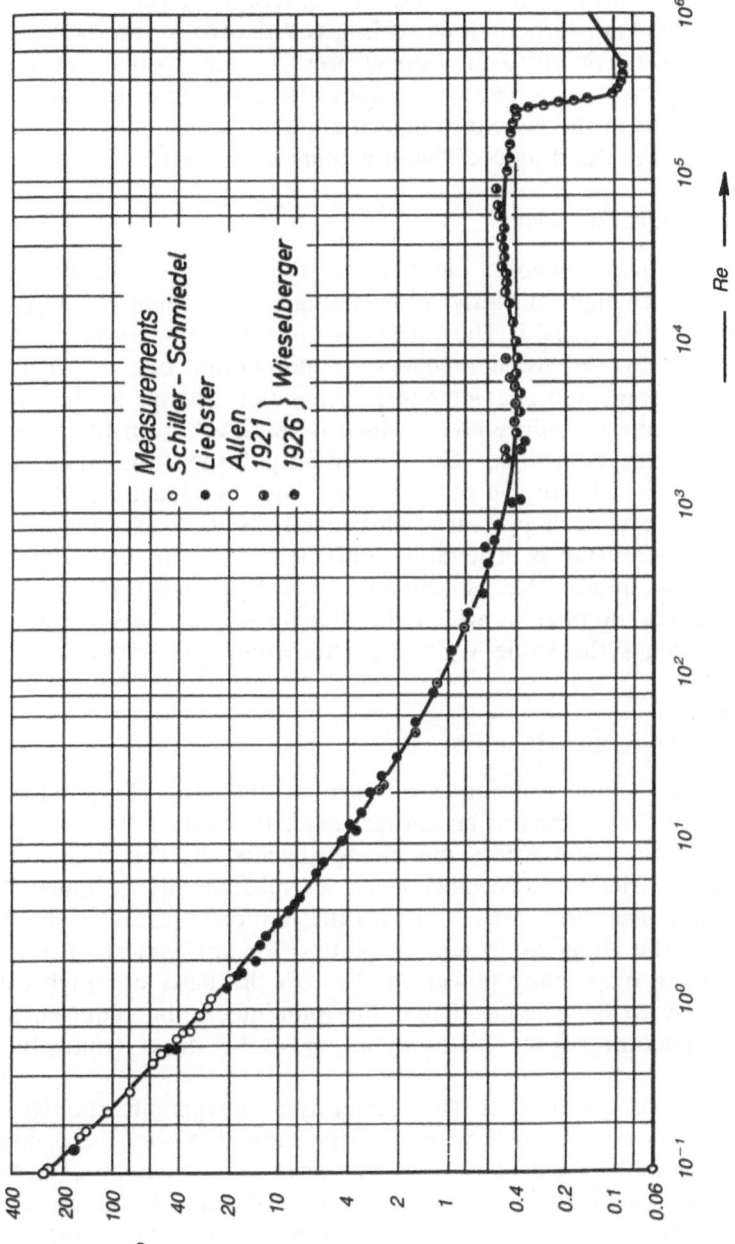

Figure 3.2 The drag coefficient for spherical particles (after Schlichting [22]).

occurs just before the equator of the sphere. The pressure in the region of separation filled with eddies behind the sphere is lower than the static pressure of the undisturbed flow. This gives rise to a resultant pressure force that is proportional to the impact pressure, i.e. to the square of the velocity of approach, as long as the extent of the region of separation does not change. Thus in the Newton range the law of resistance takes the form $C_D = 0.4$–0.5, i.e. the drag coefficient is almost constant.

(iv) Critical range: Re $\approx$ Re$_c$

Around the critical Reynolds number Re$_c$ the nature of the flow in the boundary layer changes abruptly from laminar to turbulent just before the separation point is reached. The turbulence in the boundary layer leads to an essentially more intensive exchange of momentum between it and the surrounding stream, and it gains energy from this stream. In this way the reverse flow in the boundary layer, which is responsible for the separation of the fluid stream, is avoided. The separation point is shifted downstream. As a result of this the region of separation becomes much smaller with a simultaneous recovery of pressure. This manifests itself as a marked drop in the pressure resistance. The drag coefficient drops to $C_D \approx 0.09$. The critical Reynolds number Re$_c$ is defined, following a suggestion by Dryden, as that Reynolds number at which the steeply sloping curve of the drag coefficient reaches the value $C_D = 0.3$. According to Dryden's measurements [43] Re$_c$ = 3×10^5.

(v) Supercritical range: Re > Re$_c$

If the Reynolds number is increased further, then the abrupt change in boundary-layer flow occurs at an earlier stage. In the turbulent part of this flow, between the point where this change occurs and that where separation occurs, the frictional resistance increases substantially, so much so that it becomes important in comparison with the reduced pressure resistance.

In Fig. 3.2 the drag coefficient is plotted against values of Reynolds number ranging over many powers of 10. On the basis of rough calculations it is easy to be convinced that this plot meets the requirements of chemical engineering. Reynolds numbers above 10^4 occur relatively infrequently.

In the case of accelerating or decelerating movements the Reynolds number can vary continuously within a wide range between zero and some maximum value. The calculation of the onward movement of spherical particles by the integration of the equations of motion is therefore made much easier if the plot of the drag coefficient can be approximately represented by simple easily integrated functions. Numerous approximations have been suggested for this purpose [60]. An approximation that is frequently chosen is the function

$$C_D(\text{Re}) = \frac{24}{\text{Re}} + 0.5 \qquad (3.16)$$

This is simple and can be applied over the whole range of Reynolds numbers up to Re = 10^5, Much more exact, although also more difficult to manipulate, is the approximating function

$$C_D(\text{Re}) = \frac{24}{\text{Re}} + 4\sqrt{\text{Re}} + 0.4 \qquad (3.17)$$

that has been suggested by Kaskas and Brauer [3].

(c) The influence of turbulence in the bulk of the fluid and of surface roughness

It was initially assumed that the flow of the bulk of the fluid was completely laminar. However, this does not correspond to the actual situation under most engineering circumstances. As a general rule a turbulent fluctuating velocity v' is superimposed on the mean velocity v of the fluid. Freely mobile particles follow this fluctuating movement according to the frequencies ω_f of the fluctuations, which are distributed over a wide band. If, however, $\omega_f w_s/g \gg 1$, where w_s is the settling velocity of the particles and g is the acceleration due to gravity, the particles themselves are not affected by the fluctuations, but the drag force acting on them can be substantially altered by such turbulence. The physical cause of this effect, which is observed only in the range from medium to high Reynolds numbers, is that, as a result of the turbulence in the surrounding stream, the sudden change from laminar to turbulent flow in the boundary layer occurs at an earlier stage. As a consequence of this, the characteristic drop in the drag coefficient curve is displaced towards lower Reynolds numbers, i.e. the critical Reynolds number is reduced.

These relations were investigated more closely by Torobin and Gauvin [94] for freely moving particles. The critical variable was found to be the degree of turbulence Tu, which is related to the relative velocity of approach v_{rel} and defined by the expression

$$\text{Tu} = (\overline{v'^2})^{1/2}/v_{rel} \qquad (3.18)$$

where $(\overline{v'^2})^{1/2}$ is the root mean square of the fluctuating turbulent velocity v'. The data for the range investigated can be represented by the equation

$$\text{Tu}^2\text{Re}_c = 45 \qquad (3.19)$$

Figure 3.3 shows plots of the data and smoothed curves for moving

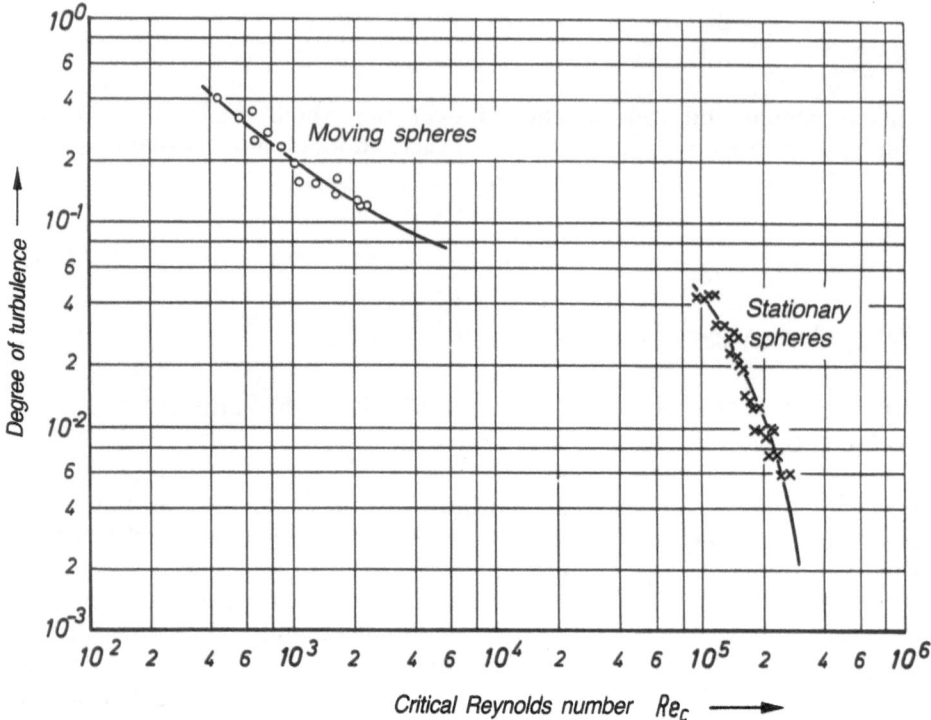

Figure 3.3 The influence of the degree of turbulence on the critical Reynolds number.

spheres. The results obtained by Dryden *et al.* [43] for stationary spheres in a turbulent stream are plotted in the same figure. In summary it can be stated that the critical Reynolds number for both moving and stationary spheres is reduced by the turbulence in the fluid flowing past them to an extent that depends on the degree of that turbulence. Before the critical Reynolds number is reached the influence of the fluid's turbulence on the drag coefficient appears to be small. However, there are as yet no reliable data for this.

The surface roughness of the particles has an effect similar to that of turbulence in the bulk of the fluid. Likewise it promotes the change in the boundary layer from laminar to turbulent flow. The explanation for this is that the asperities protruding into the surrounding fluid produce eddies. Here the relevant factor is obviously the height k_s of the asperity relative to the particle diameter x_p. Figure 3.4 shows the data of Sawatzki [21], which reveal a clear decrease in the critical Reynolds number Re_c with increasing relative roughness k_s/x_p.

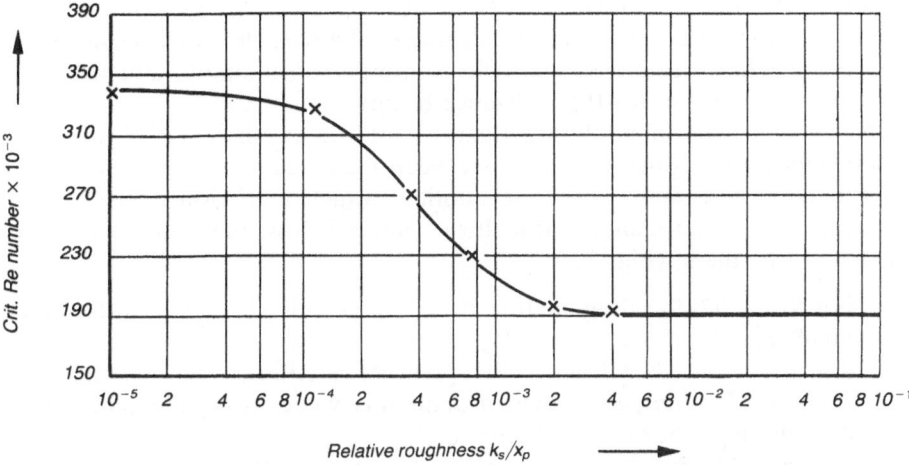

Figure 3.4 The influence of the relative roughness on the critical Reynolds number [21].

(d) The influence of compressibility and the mean free path of the fluid

The assumptions made initially that the fluid was incompressible and behaved like a continuum (mean free path several orders of magnitude smaller than the particle size) can be upheld to only a very limited extent for gases.

The influence of compressibility on the resistance of a particle to flow is usually characterized by means of the Mach number $\mathrm{Ma} = v_{\mathrm{rel}}/v_{\mathrm{s}}$, where v_{rel} stands for the velocity of approach to the particle and v_{s} is the natural velocity of propagation of small pressure disturbances in the fluid, i.e. the velocity of sound in the fluid. This velocity is given by the relation

$$v_{\mathrm{s}}^{2} = \left(\frac{\partial p}{\partial \rho_{\mathrm{f}}}\right)_{\mathrm{isentrop}} = \gamma \frac{p}{\rho_{\mathrm{f}}} \tag{3.20}$$

where p is the static pressure, ρ_{f} is the density of the fluid and γ is the ratio of the specific heat capacity at constant pressure to that at constant volume. In the case of diatomic gases (e.g. air) $\gamma = 1.40$. It is known that the velocity of sound depends only on the temperature and not on the pressure. The influence of the compressibility will be smaller for smaller Mach numbers.

Incompressibility can be assumed for $\mathrm{Ma} \ll 1$. This is almost always the case for liquids. Flows in the range $\mathrm{Ma} < 1$ are called subsonic and those in the range Ma > 1 are called supersonic. In the case of supersonic flow a shock wave is set up at the upstream side of the particle. As a result of this the field of flow around the particle is altered substantially, and the same is then also expected for the resistance to flow.

Numerous experimental investigations of the dependence of the drag coefficient on the Mach number have been made. In some cases the results are in poor agreement with one another, which is obviously due to the considerable experimental difficulties. Nevertheless the following statements can be made (Fig. 3.5):

1. For Ma < 0.6 the drag coefficient increases only slightly with the Mach number;
2. The Mach number Ma = 0.6 represents a critical value because the velocity of sound is reached around this value in certain places (i.e. near the equator of the sphere);
3. Between Ma = 0.6 and Ma = 1.5 the drag coefficient increases with increasing Mach number;
4. In the range Ma > 1.5 the drag coefficient remains approximately constant.

These observations [9, 99] are valid for Re > 100. In the range of smaller Reynolds numbers the relations become more complicated, as shown in Fig. 3.6. The curves for various Mach numbers cross one another in such a way that the drag coefficient finally assumes the smallest value in the case of the largest Mach numbers.

> The reason for this is that larger Mach numbers can only be realized for smaller Reynolds numbers under conditions such that the flowing gas can no longer be regarded as a continuum. Under these circumstances the relevant

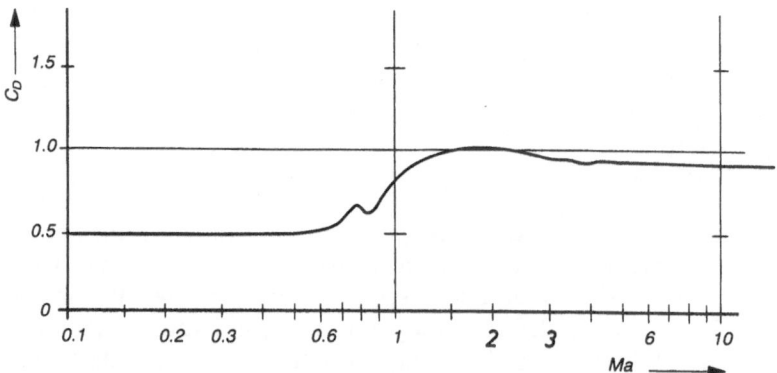

Figure 3.5 The influence of the Mach number for Re > 100 on the drag coefficient for spherical particles [9].

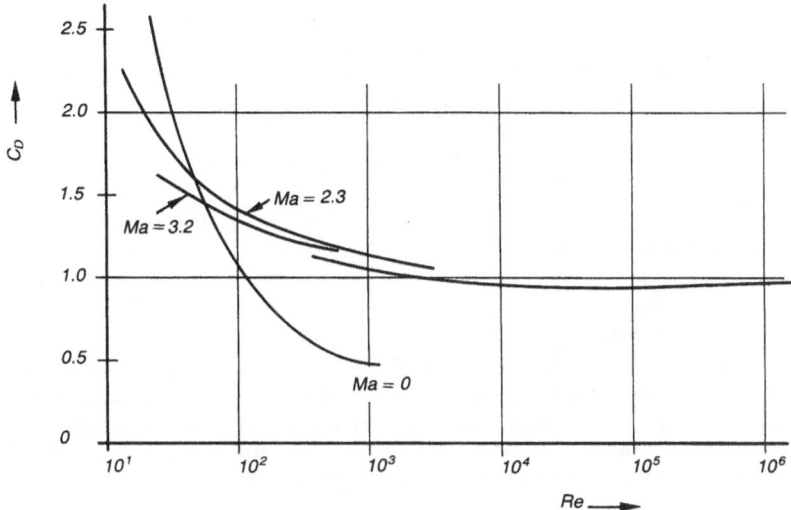

Figure 3.6 The influence of the Mach number on the drag coefficient for spherical particles in the range of large Reynolds numbers [9].

dimensionless group is the Knudsen number $\text{Kn} = \lambda/x_p$, i.e. the ratio of the mean free path λ of the gas molecules to the particle diameter x_p. However, there is a simple relation between the Knudsen number and the other two dimensionless groups. The expression

$$\text{Re Kn} = \frac{x_p v_{rel}\rho_f}{\eta}\frac{\lambda}{x_p} \tag{3.21}$$

together with the well-known equation

$$\eta = 0.499\,\overline{v_{mol}}\lambda\rho_f \tag{3.22}$$

derived from the kinetic theory of gases, where

$$\overline{v_{mol}} = \left(\frac{8p}{\pi\rho_f}\right)^{1/2} = v_s\left(\frac{8}{\pi\gamma}\right)^{1/2} \tag{3.23}$$

leads to the simple relation

$$\text{Re Kn} = \text{Ma} \times 2.004\left(\pi\frac{\gamma}{8}\right)^{1/2} \tag{3.24}$$

Here $\overline{v_{mol}}$ is the mean velocity of the gas molecules, v is the velocity of approach of the particle, v_s is the velocity of sound and ρ_f, η and γ are respectively the density, dynamic viscosity and the ratio of the specific heat capacities of the gas. If 1.40 is substituted for γ it follows that

$$\text{Re Kn} = 1.486\,\text{Ma} \tag{3.25}$$

Theoretical and experimental investigations have shown that the resistance to flow becomes smaller as the Knudsen number increases.

Davies [41] has approximated all published measurements of the drag coefficient C_D with an equation of the form

$$C_D = C_{D\,\text{Stokes}}\left\{1 + \frac{\lambda}{x_p}\left(2.514 + 0.800\exp\left(\frac{-0.55x_p}{\lambda}\right)\right)\right\}^{-1} \qquad (3.26)$$

that is valid for $0.1 < \text{Kn} < 1000$ and $\text{Re} < 0.25$. Here the mean free path of the gas molecules is defined by eqn (3.22):

$$\lambda = \frac{\eta}{0.499 v_{\text{mol}}\rho_f} \qquad (3.22a)$$

In the Davies equation it would be possible to replace the Knudsen number by the ratio of the Mach number to the Reynolds number. An expression would therefore be obtained which corresponds to Fig. 3.6 and which would supplement this figure. However, the Davies equation in the form quoted is undoubtedly to be preferred for practical applications.

(e) The influence of acceleration and particle rotation

The above discussion of the resistance of a spherical particle to flow is valid only for the assumed case of steady state flow. The flow becomes unsteady if the movement of the particle or the fluid or both together is accelerated or decelerated. In this way the resistance to flow can change substantially, and this holds not only for liquids but also for gases.

There is a theoretical solution only for the case of very small Reynolds numbers, i.e. for the so-called Stokes range. According to the calculations of Basset, Boussinesq and Oseen, which have been extended by Tchen [27] to cover the case of variable fluid velocity, the drag force D on a spherical particle of diameter x_p in unsteady flow is given by

$$D = 3\pi\eta x_p \boldsymbol{v}_{\text{rel}} + \frac{1}{2}\frac{\pi}{6}x_p^3\rho_f\dot{\boldsymbol{v}}_{\text{rel}}$$

$$+ \frac{3}{2}x_p^2(\pi\rho_f\eta)^{1/2}\int_{t_0}^{t}(t - t')^{-1/2}\dot{\boldsymbol{v}}_{\text{rel}}\,dt' \qquad (3.27)$$

Here t_0 is the time when the velocity begins to change, η and ρ_f are respectively the viscosity and density of the fluid and $\boldsymbol{v}_{\text{rel}}$ is the relative velocity between particle and fluid.

The first term on the right-hand side is the steady state drag force. The second term is a force that would occur in a completely frictionless flow. This can be explained by changes in the momentum of the flow brought about by variations in the velocity of approach $\boldsymbol{v}_{\text{rel}}$. The associated gain or loss of energy must occur by exchange with the system's surroundings. Finally, the third term is the so-called Basset resistance. This, as can be seen, is not only dependent on the change in velocity $\dot{v}_{rel}$ at the actual time t but also on its whole previous history. If the changes in velocity are very large the Basset resistance can exceed the steady state resistance.

In the range of larger Reynolds numbers it has not yet been possible to explain the influence of acceleration adequately. The data reported by various authors are contradictory, sometimes to an alarming extent. This is obviously connected with the fact that, in this range, the resistance depends not only on the actual change in the velocity at the time in question but also on its previous history. Torobin and Gauvin [93] have therefore suggested introducing the rate of change of velocity $\ddot{v}_{rel}$ as an additional parameter. This would indicate that the measured drag coefficient C_D should be evaluated according to an expression of the form

$$C_D = C_D\left(\mathrm{Re}, \frac{x_p \dot{v}_{rel}}{v_{rel}^2}, \frac{x_p^2 \ddot{v}_{rel}}{v_{rel}^3}\right) \tag{3.28}$$

This has not yet been done, not least because the accuracy of measurement is not generally sufficient to ascertain the second derivative of v_{rel}. The influence of the acceleration on the resistance in the range of larger Reynolds numbers is clearly due to changes in the separation and the wake. This is borne out by the observation that the influence of acceleration disappears at very high Reynolds numbers. According to the measurements of Torobin and Gauvin [93] turbulence in the approaching fluid also acts to suppress the effect of acceleration.

If the spherical particle around which the fluid is flowing rotates with an angular velocity ω about an axis that is perpendicular to the direction of flow, then on the one hand a dynamic thrust T_d will arise, i.e. a force perpendicular to the direction of flow will be observed, but on the other hand an increase in resistance will occur.

Figure 3.7 shows the measurements made by Sawatzki [21] in the range of subcritical Reynolds numbers. The upper curve shows the behaviour of the drag coefficient, and the lower curve shows the behaviour of the thrust coefficient which is defined in an analogous way to the drag coefficient. According to Sawatzki's data the behaviour of both curves is independent of the Reynolds number over a wide range. At very small Reynolds numbers, i.e. in the Stokes range, this state of affairs undergoes a fundamental change. Here the dynamic thrust vanishes completely and the resistance is not affected by the rotation.

(f) The influence of particle shape

If the particle is no longer spherical the relations become very much more complicated. A general theory for the Stokes range has been developed by Brenner [35], and this is described below in broad outline.

In order to move an irregularly shaped particle with a velocity w through a quiescent fluid a force $F = D + T_d$ is necessary to overcome a resistance

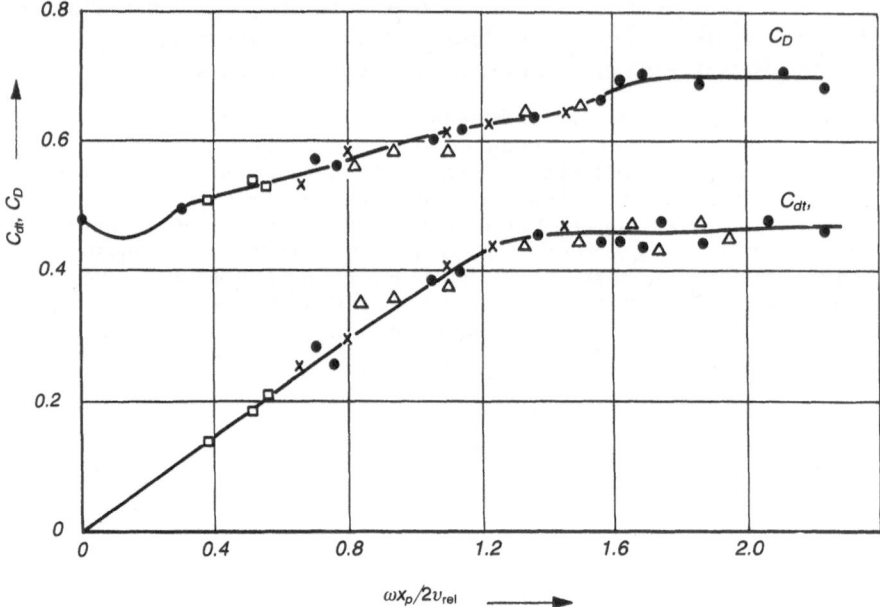

Figure 3.7 Coefficients for the dynamic thrust and drag on a smooth rotating sphere as determined by Sawatzki [21] in the subcritical range of Reynolds numbers.

D in the direction of motion plus a dynamic thrust T_d perpendicular to *w*, i.e. a force whose direction does not generally correspond to that of the velocity *w* of the particle. In addition, a torque *M* acts on the particle; this torque vanishes only when the shape of the particle fulfils certain conditions of symmetry. It follows by analogy that, when a particle of irregular shape is rotating, a torque *M* must be overcome whose line of action generally no longer corresponds with that of the angular velocity *ω*. In addition, a force *F*, which again vanishes only when certain conditions of symmetry are fulfilled, also acts on the particle.

When there is a combination of translation and rotation a force *F* and a torque *M* act on the particle and each of these depends in the most general case on both the linear velocity *w* and the angular velocity *ω*. This relation can be expressed by the tensor equations

$$F_i = -\eta(A_{ij}w_j + B_{ij}\omega_j) \tag{3.29a}$$

$$M_i = -\eta(C_{ij}w_j + D_{ij}\omega_j) \tag{3.29b}$$

where η is the dynamic viscosity of the fluid and A_{ij}, B_{ij}, C_{ij} and D_{ij} are second-order tensors which depend only on the size and shape of the particle. Brenner was able to show that A_{ij} and D_{ij} are symmetrical tensors and that B_{ij} is equal to C_{ij}.

The following important conclusions can be drawn from these general tensor equations.

1. The tensor property signifies independence from the coordinate system. If the tensor components A_{ij}, B_{ij}, C_{ij} and D_{ij} are known for a specified system of coordinates associated with the particle, the tensor components for any other system of coordinates can be easily calculated;

2. In the most general case the number of tensor components is $2 \times 6 + 9 = 21$. Thus there are 21 parameters specific to the particle that unequivocally determine the relation between movements and forces. It is therefore hopeless to expect to be able to give a complete description of the law of motion for an irregularly shaped particle in the Stokes range by means of a size parameter δ and a single additional shape parameter, as has been attempted again and again;

3. The tensor equations are linear. Movements and forces may therefore be superimposed in any way. Of course, this holds only for the Stokes range;

4. As it settles, a particle of irregular shape executes a complicated movement that varies constantly. Superimposed on a translational velocity w that varies in magnitude and direction is an angular velocity ω that varies likewise. This follows from the equilibrium of the forces acting on the particle. The force F exerted by the fluid on the particle must at any given time be equal in magnitude to the particle's weight less its buoyancy, i.e. to a force F^*. The forces F and F^* have the same direction but not the same line of action. The resultant turning movement is balanced by the torque M;

5. The tensors B_{ij} and C_{ij} vanish for particles with three mutually perpendicular axes of symmetry. The centre of gravity of such a particle lies at the point of intersection of these three axes and the particle settles with constant velocity w in a direction that deviates from the vertical by a specific angle. The size of this angle depends on the shape and orientation of the particle. Rotation of the particle is not possible, i.e. it settles in its initial attitude.

In the case of larger Reynolds numbers, i.e. in the intermediate and Newton ranges, the relation between movements and forces can no longer be described by linear tensor equations. Because of the extreme difficulties encountered a general theory for the kinetic behaviour of irregularly shaped particles for larger Reynolds numbers has not yet been formulated. It has often been observed in the sedimentation of symmetrical particles that they present their largest cross-section perpendicular to the direction of their movement relative to the fluid, i.e. the stable orientation that they assume results in the lowest possible settling velocity. Furthermore, it has been observed that in the intermediate and Newton ranges they can, under certain circumstances, execute considerable lateral and oscillatory movements. This is a consequence of the asymmetrical separation of eddies which, even with spherical particles, can manifest itself in the form of lateral oscillations.

(g) Wall effects

The resistance of a particle to flow also depends on its distance from the boundary walls within which the fluid is confined. On the assumption that the flow is viscous, theoretical solutions have been found for a series of simple geometries. The influence of a plane wall on the resistance of a sphere was calculated by Lorentz in 1896. According to him the drag coefficient C_D for a sphere of diameter x_p at a distance h from a plane horizontal wall is given by

$$C_D = \frac{24}{Re} \left(1 - \frac{9x_p}{16h}\right)^{-1}$$

(3.30a)

and for a plane vertical wall

$$C_D = \frac{24}{Re} \left(1 - \frac{9x_p}{32h}\right)^{-1}$$

(3.30b)

i.e. the drag on a spherical particle is increased by the wall effect. It has been assumed here that the plane wall is moving with the velocity v of the bulk of the fluid or, in other words, that a relative velocity v_{rel} prevails between the wall and the particle. Furthermore, it has been assumed in the derivation of this and the following equations that the diameter of the sphere is small compared with the distance h from the wall.

Ladenburg [62] obtained the following expression for the drag coefficient C_D of a sphere moving along the axis of an infinitely long cylinder of diameter d:

$$C_D = \frac{24}{Re} \left(1 + 2.104 \frac{x_p}{d}\right)$$

(3.31)

This solution has been extended by Brenner and Happel [36] to the case where the sphere is moving at any distance r from the axis of the cylinder. Their solution takes the form

$$C_D = \frac{24}{Re} \left\{1 + f(\beta) \frac{x_p}{d}\right\}$$

(3.32)

where $\beta = r/(d/2)$. The function $f(\beta)$ cannot be specified as a closed analytical expression. Brenner and Happel have provided the approximations

$$f(\beta) = 2.104 - 0.6977\beta^2 \qquad \text{for } \beta \to 0 \qquad\qquad (3.32a)$$

$$f(\beta) = \frac{9}{16}(1 - \beta)^{-1} \qquad \text{for } \beta \to 1 \qquad\qquad (3.32b)$$

Both these approximate solutions agree in the limiting case with those of Ladenburg and Lorentz.

Figure 3.8 shows the exact behaviour of the function $f(\beta)$. It is noteworthy that at $\beta = 0.40$ this function has a minimum that falls about 3% lower than the value for the axis of the cylinder ($\beta = 0$). The behaviour of $f(\beta)$ has been confirmed by measurements made by Koglin [13].

3.1.2 The motion of a single particle

(a) The equations of motion

As already pointed out in section 3.1.1(a) the following types of force can act on a single particle in a flowing medium: field forces, aerodynamic and hydrodynamic forces, pressure forces, inertial forces, diffusion forces and contact forces. In the following we shall completely ignore the diffusion and contact forces and, of the possible field forces, take into account only the gravitational force. Then, in accordance with d'Alembert's principle the gravitational force G, the hydrodynamic or aerodynamic forces F, the

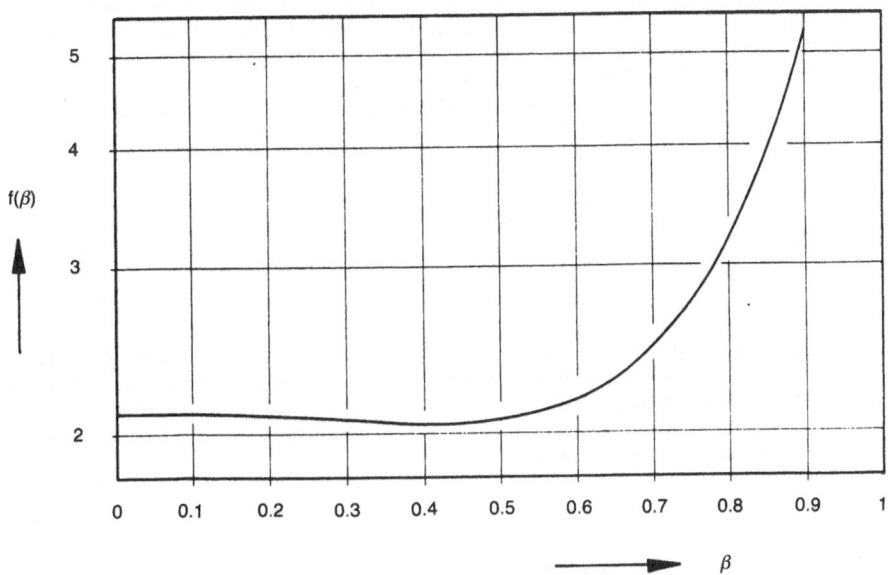

Figure 3.8 Plot of the function $f(\beta)$ according to Brenner and Happel [36].

pressure forces B and the inertial forces I constitute an equilibrium system that can be represented in vector notation by the equation

$$G + F + B + I = 0 \qquad (3.33)$$

In addition to this there should be a corresponding equation describing the equilibrium of the moments acting on the particle and one which would provide the basis for determining the rotational movement of the particle. However, the situation is such that neither the particle rotations caused by collisions with a wall nor those determined by the irregular shape of the particles can be sufficiently well predicted in most cases. That is why, when calculating the motions of particles, rotation and dynamic thrust are usually completely neglected. Of course, under these conditions we cannot expect any agreement with reality. Nevertheless the calculations are indispensable if we are to learn something about the quantitative tendencies.

The neglect of the dynamic thrust means that in the above force balance the aerodynamic (or hydrodynamic) force F can be substituted by the drag force D. This is given by the relation

$$D = C_D A_p \frac{\rho_f}{2} v_{rel} \boldsymbol{v}_{rel} \qquad (3.34)$$

Here ρ_f is the density of the fluid, $\boldsymbol{v}_{rel}$ is the relative velocity between particle and fluid, A_p is the projected area of the particle in the direction of flow and C_D is the drag coefficient which, as has been shown above, can depend on many variables other than the Reynolds number. How many and which of these variables must be taken into account in each case is determined by the problem in question.

The following expressions apply for the weight G, the body force B and the inertial force I:

$$G = mg \qquad (3.35a)$$

$$B = -V \operatorname{grad} p \qquad (3.35b)$$

$$I = -m\dot{w} \qquad (3.35c)$$

where m and V are respectively the mass and volume of the particle, g is the acceleration due to gravity, w is the translational velocity of the particle and $\operatorname{grad} p$ is the pressure gradient in the field of flow. If we restrict ourselves to non-dissipative flows, i.e. flows for which $\eta \nabla^2 v$ vanishes, then it follows from the Navier–Stokes equations that

$$\operatorname{grad} p = \rho_f(g - \dot{v}) \qquad (3.5a)$$

where v stands for the velocity of the fluid.

If all these expressions are substituted in the force balance given initially, it follows that

$$\frac{\mathrm{d}\mathbf{w}}{\mathrm{d}t} = \frac{C_\mathrm{D}A_\mathrm{p}\rho_\mathrm{f}}{2m} \, v_\mathrm{rel}\mathbf{v}_\mathrm{rel} + \frac{(\rho_\mathrm{p} - \rho_\mathrm{f})\mathbf{g}}{\rho_\mathrm{p}} \frac{\rho_\mathrm{f}}{\rho_\mathrm{p}} \frac{\mathrm{d}\mathbf{v}}{\mathrm{d}t} \tag{3.36}$$

where ρ_p is the density of the particle. If the relation

$$\mathbf{v}_\mathrm{rel} = \mathbf{v} - \mathbf{w}$$

is taken into account, three coupled differential equations of the first order are obtained as component equations.

(b) Uniform motion of a particle

The motion of a particle can often be divided into an initial phase during which the particle is accelerated or decelerated and a second phase during which the particle continues to move with constant velocity. For this it is assumed that the velocity $\mathbf{v}$ of the fluid itself does not vary from time to time or from place to place, i.e. $\mathrm{d}\mathbf{v}/\mathrm{d}t = 0$. If we are interested in only the second phase of the particle's motion, i.e. the phase when the translational velocity $\mathbf{w}$ of the particle is constant, then $\mathrm{d}\mathbf{w}/\mathrm{d}t$ can be set equal to zero in the vector equation of motion, which now becomes

$$\frac{C_\mathrm{D}A_\mathrm{p}\rho_\mathrm{f}}{2m} \, v_\mathrm{rel}\mathbf{v}_\mathrm{rel} + (\rho_\mathrm{p} - \rho_\mathrm{f}) \frac{\mathbf{g}}{\rho_\mathrm{p}} = 0 \tag{3.37}$$

We now solve this equation for the steady state velocity $\mathbf{v}_\mathrm{rel}$ of the particle relative to the fluid; the magnitude of this velocity is equal to that of the terminal settling velocity w_f of the particle in the quiescent fluid. Here C_D is the drag coefficient, A_p is the projected area of the particle in the direction of flow, ρ_p and m are respectively the density and mass of the particle, ρ_f is the density of the fluid and $\mathbf{g}$ is the acceleration due to gravity.

In the case of spherical particles in the Stokes range it is extremely easy to solve the equation. The solution obtained is

$$w_\mathrm{f} = \mathbf{v}_\mathrm{rel} = \frac{x_\mathrm{p}^2}{18} (\rho_\mathrm{p} - \rho_\mathrm{f}) \frac{\mathbf{g}}{\eta} \tag{3.38}$$

if no variables other than the Reynolds number have to be taken into account (x_p is the particle diameter and η is the dynamic viscosity of the fluid). At larger Reynolds numbers there is some difficulty in that in each case the drag coefficient C_D must be read off from Fig. 3.2 as a function of the Reynolds number. However, in accordance with its definition

$$\mathrm{Re} = v_\mathrm{rel}x_\mathrm{p} \frac{\rho_\mathrm{f}}{\eta} \tag{3.39}$$

the Reynolds number contains the quantity being sought, i.e. v_{rel}. In any case an exact solution can be achieved relatively quickly by iteration. The effort involved can be substantially reduced by using the diagram devised by Grassmann [47].

An approximately uniform particle movement can also be established in centrifuges if the fluid rotates like a solid body with the casing of the centrifuge. If $\mathbf{\Omega}$ is the angular velocity of the centrifuge, $\mathbf{R}$ is the radius vector at the location of the particle and $\mathbf{w}^*$ is the velocity of the particle relative to a co-rotating system of coordinates, then the translational velocity $\mathbf{w}$ of the particle is given by

$$\mathbf{w} = \mathbf{w}^* - [\mathbf{R} \times \mathbf{\Omega}] \tag{3.40}$$

and

$$\frac{d\mathbf{w}}{dt} = \frac{d\mathbf{w}^*}{dt} - 2[\mathbf{w}^* \times \mathbf{\Omega}] - \Omega^2\mathbf{R} \tag{3.41}$$

The first term on the right-hand side of eqn (3.41) is the relative acceleration of the particle, the second is the Coriolis acceleration and the third is the centripetal acceleration. If we assume that the relative particle velocity $\mathbf{w}^*$ is small compared with the circumferential velocity $-[\mathbf{R} \times \mathbf{\Omega}]$, then it can readily be shown that the first two terms can be neglected in comparison with the third, i.e.

$$\frac{d\mathbf{w}}{dt} = -\Omega^2\mathbf{R} \tag{3.42}$$

For the fluid

$$\frac{d\mathbf{v}}{dt} = -\Omega^2\mathbf{R} \tag{3.43}$$

If these two expressions are substituted in the equation of motion for the particle it follows that

$$\frac{C_D A_p \rho_f}{2m} v_{rel} \mathbf{v}_{rel} + \frac{\rho_p - \rho_f}{\rho_p} (\mathbf{g} + \Omega^2\mathbf{R}) = 0 \tag{3.44}$$

Here the acceleration due to gravity can usually be neglected in comparison with the centripetal acceleration. Hence we have obtained an equation for the settling velocity in a centrifuge which has the same form as the equation for the settling velocity in the gravitational field. Nevertheless there is a distinct difference between the two equations: the centrifugal force is not constant like the gravitational force but varies with the radius of the centrifuge. The motion of the particle in the centrifuge can therefore be designated only as quasi-steady state.

The assumption that the relative velocity w^* of the particle is small compared with the circumferential velocity is valid for most centrifugal separating processes, and in any case for the particle size range for which centrifugal separations are used.

(c) Accelerated particle movement

In the case of accelerated particle movement the complete equation of motion (3.36) must be used. As already mentioned, the component equations thus obtained are three coupled differential equations of the first order. Only in exceptional cases is there an analytical solution for this system of equations. These exceptions are as follows.

1. Particle movements in the Stokes range: in the Stokes range the drag force is directly proportional to the relative velocity v_{rel} so that the coupling no longer applies and the equations can be solved individually;
2. Purely vertical movements: if the particles move in the fluid exclusively in the vertical direction then all the terms in the vectorial equation of motion have the same direction and it can be written as a scalar equation. The vectorial equation of motion has, as it were, only one component equation. This can always be solved if an integrable approximating function is introduced for the dependence of the drag coefficient C_D on the Reynolds number Re;
3. Movements in a flow field of constant velocity where gravity is neglected.

In a flow field of constant velocity $dw = -dv_{rel}$. Substituting this in the vectorial equation of motion and eliminating the last two terms yields the result

$$\frac{dv_{rel}}{dt} = -C_D A_p \frac{\rho_f}{2m} v_{rel} v_{rel} \tag{3.45}$$

This equation states that the velocity change dv_{rel} occurs along the same line of action as the velocity v_{rel} itself, i.e. the relative velocity can change in magnitude but not in direction. This is illustrated in Fig. 3.9. The directions of the fluid velocity v and the relative velocity v_{rel} remain fixed. Only the direction of the particle velocity is variable. The relative velocity decreases continuously from an initial maximum value down to zero. As a result the vector of the particle velocity approaches the vector of the fluid velocity more and more closely until it coincides with it. Since all terms of the equation of motion have the same direction it can be written in scalar form. The integration is then just as simple as in the case of the purely vertical movement [60].

Apart from these exceptional cases, the equations of motion can only be solved using stepwise numerical integration. Such solutions have frequently been developed for special technically significant types of flow [33, 69, 103].

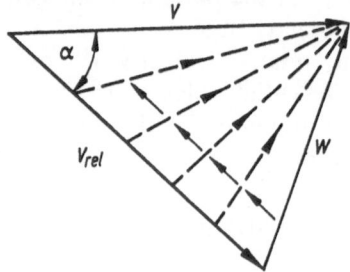

Figure 3.9 Change in particle velocity with time in a flow field of constant velocity where gravity is neglected.

3.1.3 The dynamic interaction of particles in a fluid

(a) The interaction between individual particles

In the case of creeping flow the mutual interaction of two spherical particles of equal size was dealt with theoretically by Smoluchowski in 1911 [88]. He then made use of the reflection method which had first been used by Lorentz to determine the wall effect. This method is mathematically simple but can only be applied when the distance between the particles is large compared with the particle diameter x_p.

Smoluchowski found that in every case the resistance to flow of a single spherical particle is reduced by the proximity of a second such particle. Its settling velocity is increased to the same extent, i.e. in every case two neighbouring particles settle faster than a single particle. If the particles are aligned one behind the other in the direction of flow then it follows from Smoluchowski's approximate solution that the drag coefficient for each of the two particles is given by

$$C_D = C_{D\,\text{Stokes}} \left\{ 1 - \frac{3}{4}\frac{x_p}{l} + \left(\frac{3}{4}\frac{x_p}{l}\right)^2 \right\} \qquad (3.46)$$

If they are aligned beside each other across the direction of flow then the result is

$$C_D = C_{D\,\text{Stokes}} \left\{ 1 - \frac{3}{8}\frac{x_p}{l} + \left(\frac{3}{8}\frac{x_p}{l}\right)^2 \right\} \qquad (3.47)$$

Using these two equations it is possible to obtain the solution for any given direction of the approaching flow by simple superposition.

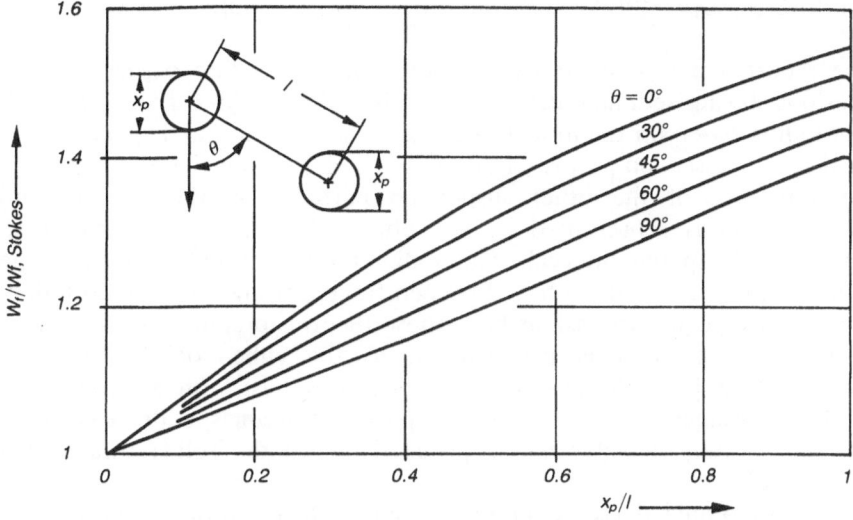

Figure 3.10 Settling velocity for two identical spheres (after Goldman *et al.* [46]).

In addition to Smoluchowski's solutions by approximate methods exact solutions that apply to particles separated by any given distance have been given more recently [46, 90]. Figure 3.10 gives a graphical representation of the results of the very complicated calculations. The ratio of the settling velocity w_f of two neighbouring particles to the settling velocity $w_{f\,Stokes}$ of a single spherical particle is plotted against the ratio x_p/l with the angle θ defining the alignment as a parameter. These theoretical results can be confirmed experimentally in the range of very small Reynolds numbers.

For three or more neighbouring spheres approximate solutions by the reflection method have been obtained only for special alignments [51, 61]. This situation generally results in relative movements between the particles. The settling velocity of complexes consisting of up to ten similar particles, whose alignment and relative movement is developing during sedimentation, has been experimentally investigated. According to these experiments the settling velocity for complexes of spheres is roughly proportional to the square root of the number of particles in the complex [56]. For other particle shapes the settling velocity of complexes is somewhat greater than the settling velocity of a single particle [15].

In the range of higher Reynolds numbers there are no theoretical solutions for the interaction between neighbouring particles. The few known experimental investigations show that in this range the resistances of two equally large particles aligned behind each other are no longer equal, with the result that during sedimentation the trailing particle catches up with the one in front.

(b) Interactions between particles settling in suspensions

The theoretical treatments of hydrodynamic interaction occurring in suspensions during sedimentation can be divided into three groups [32]. The first [50] is based on the model of a regular geometrical arrangement of the particles. The second [40] is based on a cell model in which the hydrodynamic influence of the other suspended particles on the particle under consideration is represented by the effect of a solid wall located at a characteristic spatial interval. The sedimentation of polydisperse suspensions of spherical particles has also been treated using this method [10, 34]. For monodisperse suspensions both these theories predict a decrease in the settling velocity that is proportional to the cube root of the solids concentration [45]. A third group of theories based on a geometrically random arrangement of the settling particles predicts a decrease in the average settling velocity that is proportional to the solids concentration [32, 38, 76].

The monotonic decrease in the settling velocity with increasing solids concentration C_v has been experimentally confirmed in the range of higher concentrations above $C_v \approx 10^{-2}$. However, measurements within suspensions of low concentration first show an increase in the average settling velocity with increasing solids concentration [11, 54]. This is attributable to the faster collective settling of clusters consisting of several particles. Because of the statistical distribution of the sizes of these clusters the settling velocities, even in monodisperse systems, also conform to a statistical distribution; this distribution can be shown to be log-normal [55].

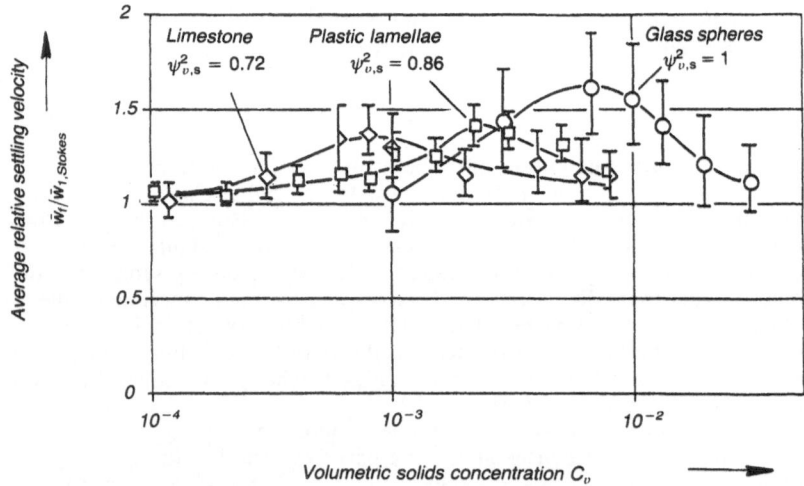

Figure 3.11 Settling velocity in monodisperse suspensions of low concentration for particles of various shapes: shape factor $\psi_{v,s}^2 = (x_v/x_s)^2$ (see p. 14).

The distribution of settling velocities in previously thoroughly mixed suspensions reaches a dynamic equilibrium after a settling path of several hundred particle diameters [15]. The ratio of the settling velocities to the Stokes velocity depends on the size, shape and density distributions of the solid as well as on the effect of vessel walls [13, 14, 31, 81]. Figure 3.11 shows the equilibrium value of the average settling velocity as a function of the volumetric solids concentration for monodisperse suspensions of glass spheres, plastic lamellae and limestone particles.

> The discrepancy between the measured results, which are usually obtained for suspensions of particles larger than 100 μm at Reynolds numbers Re $\ll$ 1, and the theoretical results based on either a regular or spatially random arrangement of particles can be attributed to the fact that during sedimentation a demixing of the suspension occurs by the formation and growth of particle clusters that settle together. A simpler example that has been demonstrated both theoretically and experimentally is the breaking up of a horizontal row of similar spheres into particle clusters separated from one another [39]. There is a lack of more precise investigations of the corresponding process in suspensions.

3.2 FLOW THROUGH PACKINGS

Flow through porous systems is a basic process that very frequently occurs in chemical engineering processes and other technologies as well as in nature. Examples include the drying of bulk solids, the filtering of liquids and gases, flow through packed columns, shaft furnaces and beds of catalysts and adsorbents, fluidization, the measurement of surface area, and the flow of ground waters and crude oil through soil and earth.

The packings through which the fluid flows can be either fixed or mobile. This does not matter as far as the dimensional analysis of the problem is concerned but it must be taken into account when dealing with the experimental laws of resistance. When the flowing medium is a gas the ratio of the mean free path λ to a characteristic pore width l_p enters into the argument. Accordingly a distinction is made between Knudsen's molecular flow when $\lambda/l_p \gg 1$, the non-slip flow of a continuum without wall adhesion when $\lambda/l_p \approx 1$ and the normal adhering flow of a continuum when $\lambda/l_p \ll 1$. Catalysts and adsorbents often have very fine pores and possess a large internal surface. The penetration of gas and liquid molecules into these systems, or rather their movement in the pore system as a consequence of concentration gradients, is known as pore diffusion. Pore diffusion is flow through pores due to the effect of partial pressures and generally occurs in the range of Knudsen or sliding flow.

It is also necessary to distinguish between single-phase and multiphase flow. A multiphase flow of fluid and particles occurs in bed filtration; in

Section 3.2 was written with the cooperation of M. H. Pahl.

the packing of the filter the particles are removed from the stream and deposited on the surface of the packing. The movement of liquid and vapour in the pore space of a material undergoing drying is one example of a liquid–gas multiphase flow; another is the removal of residual moisture in a filter cake by 'dry blowing'.

3.2.1 Dimensional analysis

Let us consider the steady isothermal single-phase flow of a continuum. The medium is assumed to be incompressible and Newtonian and hence completely specified by its density ρ_f and viscosity η. Of interest is the drop in pressure $\Delta p = p_1 - p_2$ along the length L of the packing as a function of the velocity. The relevant velocity is the superficial velocity $\bar{v}$ defined as the volumetric flow rate $\dot{V}$ divided by the free cross-sectional area A of the straight conduit filled with the packing: $\bar{v} = \dot{V}/A$. Wall and entry effects are neglected.

The difficulties in comprehending the perfusion exactly relate to the designation of the packing and the measurement of its characteristic properties. Here we shall consider only the simple proposition that there is a random arrangement of geometrically similar particles and that the particle assemblies under consideration have similar distributions. With these assumptions dimensional analysis leads to the expression [83]

$$\text{Eu} = \phi\left(\varepsilon, \frac{L}{\bar{x}}, \text{Re}\right) \tag{3.48}$$

where Eu and Re are the Euler and Reynolds numbers respectively and are defined as $\text{Eu} = \Delta p / \rho_f \bar{v}^2$ and $\text{Re} = \bar{v}\bar{x}\rho_f/\eta$, ε is the porosity of the packing and $\bar{x}$ is the location parameter of the particle size distribution.

If the porous system does not vary in the L direction apart from local statistical variations – small domains in comparison with the length of the packing – we then have quasi-homogeneous packing. For this the pressure gradient in the direction of flow is constant. Hence it follows that

$$\text{Eu} = \frac{L}{\bar{x}} F(\text{Re}, \varepsilon) \tag{3.48a}$$

If, furthermore, the function $F(\text{Re}, \varepsilon)$ can be written as the product of two single-parameter functions, this then leads to the final expression

$$\text{Eu} = \frac{L}{\bar{x}} f_1(\text{Re}) f_2(\varepsilon) \tag{3.48b}$$

If a porosity function of the form [68]

$$f_2(\varepsilon) = \frac{a(1 - \varepsilon)}{\varepsilon^b} \qquad (3.48c)$$

is chosen, the result obtained by dimensional analysis for $\varepsilon \to 1$ thus becomes the same as that for the flow around a single sphere, and it can also be applied, with a small correction to the constants, to the fluidized bed in which the porosity ε can be greater than 0.9.

3.2.2 Empirical perfusion laws

Gupte's investigations [7] of uniform random packings of spheres of almost equal size in the porosity range $0.37 < \varepsilon < 0.64$ yielded as the best fit for $Re < 1$

$$\mathrm{Eu} = \frac{L}{\bar{x}} \frac{22.4}{\mathrm{Re}} \frac{1 - \varepsilon}{\varepsilon^{4.55}} \qquad (3.49)$$

In Fig. 3.12 the results are presented in the form

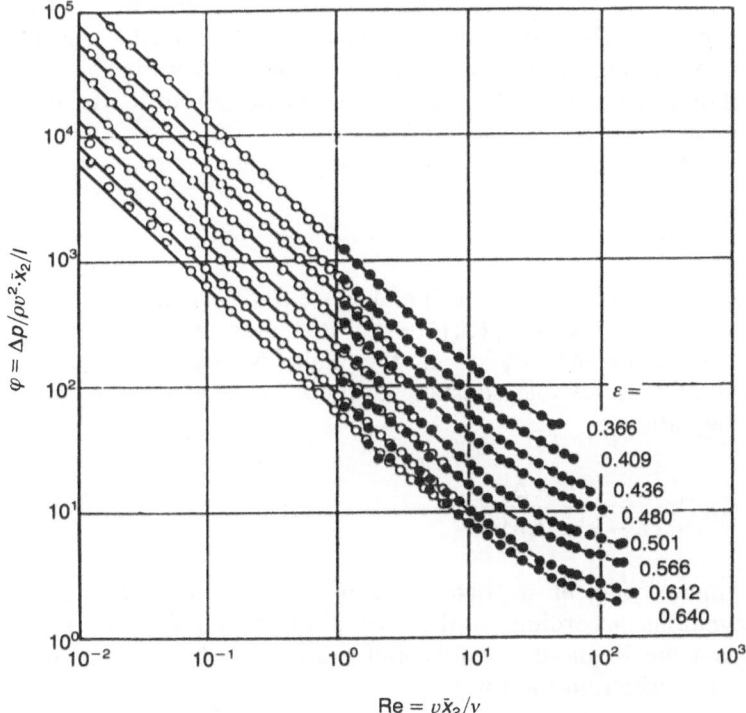

Figure 3.12 Resistance coefficient $\phi = \mathrm{Eu}(\bar{x}/L)$ for packings of spheres of equal size [7]. Fluid medium: $\bigcirc$, oil; $\bullet$, air.

$$\mathrm{Eu}\,\frac{\bar{x}}{L} = \frac{\Delta p}{\rho_f \bar{v}^2}\,\frac{\bar{x}}{L} \tag{3.50}$$

as a function of $\mathrm{Re} = \bar{v}\bar{x}\rho_f/\eta$ with ε as the parameter. The moment $M_{1,2} = \bar{x}$, which is inversely proportional to the volume-related specific surface $S_v \approx M_{2,0}/M_{3,0}$, has been substituted for the location parameter $\bar{x}$.

The first formulation of the perfusion law was given by Darcy [5] for the laminar range. Only η occurred as a material constant of the fluid in his formula:

$$\bar{v} = -\frac{B}{\eta}\,\frac{\Delta p}{L} \tag{3.51}$$

The constant B is called the 'permeability' and comprises all the factors that influence the resistance. Because of the still relatively poor state of knowledge of the dependence of the permeability on the particle size distribution, particle shape and packing structure, generally nothing can be done other than to measure B for specified packings if exact values are required. Kozeny [57] and Carman [4] have ascertained the dependence of the permeability B on the particle size, or, rather, on the specific surface of the size grading, and have derived a perfusion law on the basis of the Hagen–Poiseuille relation for the average velocity v of a stream in a tube of diameter d:

$$v = \frac{d^2 \Delta p}{32 \eta L} \tag{3.52}$$

The packing is considered to be a system of channels whose average hydraulic diameter $d_h = \varepsilon/\{S_v(1 - \varepsilon)\}$ is a measure of their size and in which the mean velocity v of the fluid is equal to $\bar{v}/\varepsilon$. With the introduction of a suitable constant k, these concepts lead to the Carman–Kozeny equation

$$\bar{v} = k^{-1}\,\frac{\varepsilon^3}{(1 - \varepsilon)^2}\,\frac{\Delta p}{L \eta S_v^2} \tag{3.53}$$

Measurements yield an average value of $k = 5$, admittedly with considerable deviations according to the size grading and type of packing. By introducing the Reynolds number and substituting $6/\bar{x}$ for S_v the equation can then be written in the form

$$\mathrm{Eu} = \frac{L}{\bar{x}}\,\frac{180}{\mathrm{Re}}\,\frac{(1 - \varepsilon)^2}{\varepsilon^3} \tag{3.53a}$$

If the Reynolds number is based on d_h as the characteristic linear dimension, then

$$\text{Re}_{CK} = \bar{v}d_h \frac{\rho_f}{\varepsilon\eta} = \bar{v} \frac{\rho_f}{S_v(1 - \varepsilon)\eta} \tag{3.53b}$$

For the whole range of Reynolds numbers the resistance law takes the form

$$\text{Eu} = \frac{L}{\bar{x}} \frac{(1 - \varepsilon)^2}{\varepsilon^3} \psi(\text{Re})\phi(\varepsilon) \tag{3.53c}$$

The dependence of the porosity function $\phi(\varepsilon)$ for $\text{Re} > 1$ on particle shape and size distribution is not known. $\phi(\varepsilon)$ is equated to unity. $\psi(\text{Re})$ is determined by the definition of Re. Usually Re_{CK} or $\text{Re}^* = \bar{x}\bar{v}/[(1 - \varepsilon)v] \approx \text{Re}_{CK}/6$ is chosen. $\psi(\text{Re}^*) = \text{Eu}(\bar{x}/L)\varepsilon^3/(1 - \varepsilon)$ is plotted in Fig. 3.13. According to Ergun [44] this function can be approximated by

$$\psi(\text{Re}^*) = \frac{150}{\text{Re}^*} + 1.75 \tag{3.54}$$

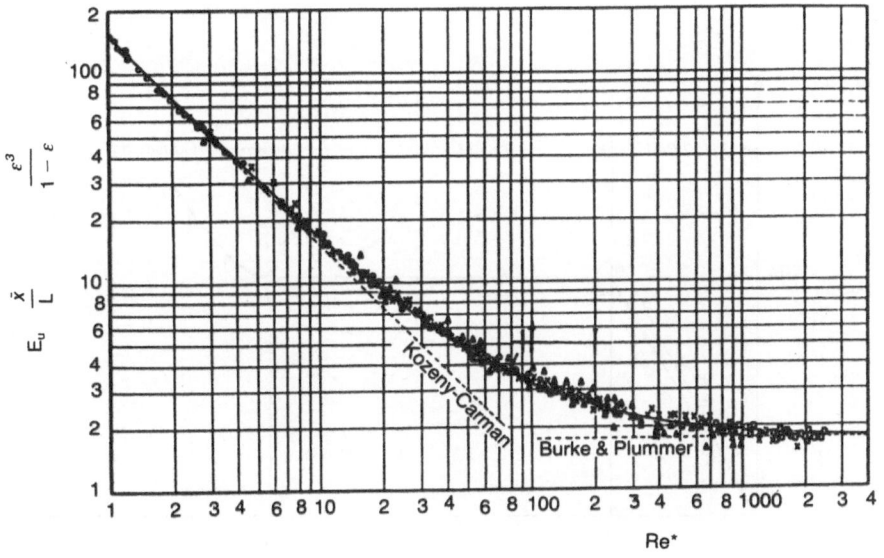

Figure 3.13 Measured values of the pressure loss for isometric particles [44].

3.3 THE MECHANICS OF THE FLUIDIZED BED

3.3.1 Phenomenology

Perfused packings, fluidized beds and pneumatically or hydraulically conveyed materials represent systems whose states merge into one another. With this in mind let us consider a poured granular material supported in a cylindrical container of cross-sectional area A on a finely perforated plate through which a fluid is flowing from below. The pressure loss Δp through the packing is a function of the velocity of the fluid which here, as previously, will be taken as the superficial velocity $\bar{v}$ defined by the formula $\bar{v} = \dot{V}/A$ where $\dot{V}$ is the volumetric velocity. The relation between Δp and $\bar{v}$ is shown in Fig. 3.14. Three distinct ranges can be seen.

When the velocity $\bar{v}$ is small the particles around which the flow is taking place form a fixed bed of constant porosity ε; the weight of the material is greater than the force exerted by the stream of fluid. The pressure loss Δp increases according to the perfusion law for fixed beds. If the minimum fluidizing velocity $\bar{v}_{mf}$ is reached the bed loosens up and all the particles begin to move around. The corresponding point on the curve is called the **fluidizing point** FP. With further increases in the velocity, Δp remains almost constant. The bed is supported by the flowing medium. Its behaviour is comparable in many respects to that of a liquid; for example, an almost flat surface is formed without any conical protrusions, and the material flows over an overflow or out of an opening on the side like a

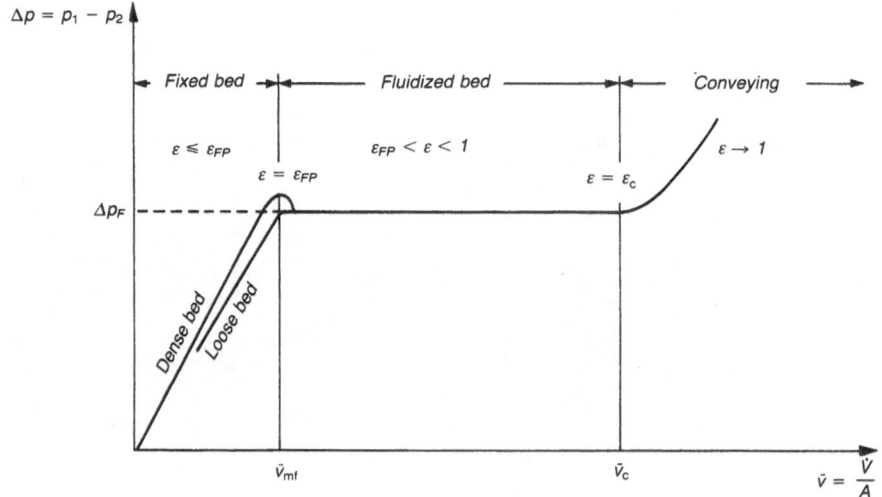

Figure 3.14 Pressure drop in fixed and fluidized beds.

this state the system is said to be fluidized, i.e. a **fluidized bed** or **fluidized layer** has been formed. The equilibrium of the forces balancing one another within the boundaries of the whole system yields the following expression for the pressure drop Δp in a fluidized bed of depth L:

$$\Delta p = \Delta p_r + L\rho_f g \tag{3.55}$$

where $\Delta p_r = (1 - \varepsilon)L\Delta\rho g$, $\Delta\rho = \rho_p - \rho_f$ is the difference between the densities of the solid and the fluid, and g is the acceleration due to gravity. Δp_r is the pressure drop caused by the resistance to flow. In the case of gases $\Delta p \approx \Delta p_r$.

At the fluidizing point there is a slight increase in Δp because an additional force is necessary to loosen up the bed. The resistance to the stream of fluid remains constant as $\bar{v}$ increases. This is a consequence of a further loosening up of the bed; its porosity increases, i.e. it expands.

If the fluid velocity in the expanded bed approaches the settling velocity of the particles, then pneumatic or hydraulic conveying sets in. Under steady operating conditions the drop in pressure again increases over the bed depth L.

The fluidized bed can be either homogeneous or inhomogeneous, i.e. it can manifest either a stable or an unstable state of motion† (Fig. 3.15). In a homogeneous fluidized bed consisting of particles of similar size and density a more or less uniform random mixture prevails; the distribution of particles and voids within the whole space is statistically similar. The

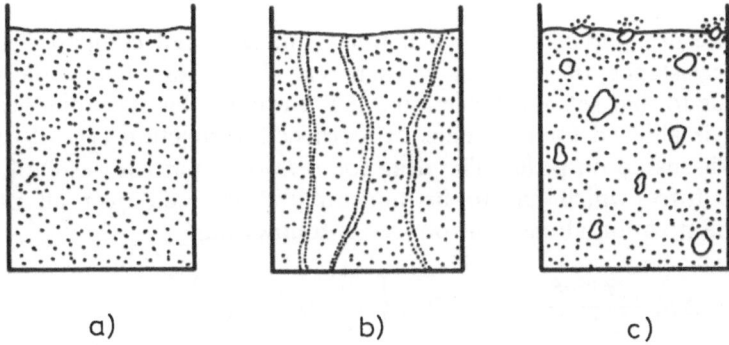

a)   b)   c)

Figure 3.15 States of a fluidized bed: (a) homogeneous fluidized bed; (b) channel formation; (c) bubble formation.

†Translator's note: it is more usual in English to speak of homogeneous fluidization as 'particulate' and inhomogeneous as 'aggregative'.

statistical state of motion is very uniform without preference for any direction and is more or less comparable with the Brownian movement. A convective circulating movement directed downwards at the walls can be superimposed. With the growth of instabilities the uniform state of motion disappears. The instabilities manifest themself as a local demixing of fluid and particles with the formation of bubbles and channels. Channel formation is found to be favoured in the case of materials whose shape deviates greatly from that of a sphere and especially at velocities around the fluidizing point. The formation of bubbles is the usual kind of inhomogeneity in a fluidized bed. As exhaustive investigations have shown, an individual bubble in the fluidized bed possesses an almost spherical shape with a pronounced indentation in its lower side (spherical-cap bubble with indented base). A thick layer of particles can build up on the surface of the bubble. This layer forms a sort of arch and thereby stabilizes the bubble. A wake is formed behind the bubble with intense axial mixing of the solid.

As a rule, of course, whole swarms of bubbles are formed. According to Werther's measurements [101] the bubble size distribution at any position in the bed is very nearly log-normal, with the average size of bubble increasing with increasing distance from the feed plate (distributor). This leads to the conclusion that a marked coalescence of the bubbles occurs as they rise. Furthermore, the bubbles show a tendency to migrate towards the centre of the bed. If the size of the bubbles approaches the diameter of the apparatus then the bed starts to slug, i.e. the bubble lifts the material above it and carries it upwards like a piston.

3.3.2 Perfusion law

The resistance per particle in a laminar fluidized bed compared with the Stokes resistance of a single particle can be determined in a manner analogous to that set out in section 3.2.1. In doing this the influence of the porosity is allowed for by means of a correction function $c(\varepsilon)$ of the form ε^{-n}, as discussed on p. 82. The drag force D on a particle results from the pressure force which has to be uniformly distributed over all particles. Using the notation already introduced it follows that

$$D = \Delta p_r \, \frac{\pi \bar{x}^3 / 6}{L(1 - \varepsilon)} = f^*(\text{Re}) \left\{ \frac{\bar{x}^2 \pi}{4} \, \rho_f \frac{\bar{v}^2}{2} \, c(\varepsilon) \right\} \tag{3.56}$$

When the relation derived previously for p_r is taken into account and $(4/3)f(\text{Re})$ is substituted for $f^*(\text{Re})$, we obtain

$$\frac{g\bar{x}}{\bar{v}^2} \frac{\Delta\rho}{\rho_f} = \varepsilon^{-n} f(\text{Re}) \qquad \text{Re} = \bar{v}\bar{x}/\nu \tag{3.56a}$$

The dimensionless expression on the left-hand side of the equation plays an important role in all two-phase flows under the influence of gravity. Its reciprocal value is called the settling coefficient Se. It indicates the ratio of the forces exerted by the flowing medium to the gravitational force acting on the phase dispersed in it. With this shorter notation the result can be written simply as $\varepsilon'' = \mathrm{Se}f(\mathrm{Re})$.

Various approximations are possible for $f(\mathrm{Re})$, e.g. $f(\mathrm{Re}) = k_F/\mathrm{Re} + C_F/\mathrm{Re}^m$; when this is used, the equation for flow through a fluidized bed takes the form

$$\mathrm{Se}\left\{\frac{k_F}{\mathrm{Re}} + \frac{C_F}{\mathrm{Re}^m}\right\} = \varepsilon^n \tag{3.57}$$

The experimental work of Lewis *et al.* [64] in the range Re < 1 yielded the results $k_F = 15\text{--}21$, $n = 4.65$ and $C_F = 0$. In the larger range of Re up to 10, Wen and Yu [100] found good agreement between their measurements and those of Lewis *et al.* and other authors by putting $f(\mathrm{Re}) = 18/\mathrm{Re} + 2.7/\mathrm{Re}^{0.313}$ and $n = 4.7$.

3.3.3 The stability of the fluidized bed

In a homogeneous fluidized bed the location of the particles, their velocity and the porosity all vary statistically at any one place with approximately the same constant probability. Small regions of higher porosity that can serve as nuclei for bubbles are continuously formed. Furthermore, inhomogeneities in the distribution of the fluid, as well as bubbles, can enter the bed from the distributor. Such inhomogeneities do not grow under stable conditions.

Molerus [67] has derived an index S as a criterion for the stability of a fluidized bed:

$$S = \frac{f(\varepsilon)}{\{\varepsilon^3(1 - \varepsilon)(\rho_p/\rho_f)\mathrm{Ar}\}^{1/2}} \tag{3.58}$$

According to this the stability is governed by the porosity and the product of the Archimedes number Ar and the density ratio ρ_p/ρ_f where $\mathrm{Ar} = (x^3g/v^2)(\Delta\rho/\rho_f)$ and x is the particle size, g is the acceleration due to gravity and v is the kinematic viscosity of the fluid.

The calculation for the stability yields a critical value S_{cr} of about 0.014. The fluidized bed is stable as long as $S > S_{cr}$ and becomes unstable for $S < S_{cr}$. The value of 0.014 for S_{cr} comes from a comparison of many data with the theory. In Fig. 3.16 the S_{cr} lines are plotted on the $(\varepsilon, \rho_p/\rho_f)$ diagram for various particle sizes. The calculation of the stability is based on an ideal model. It is not to be expected that it would quantitatively and exactly reflect actual conditions. However, the trends are in accord with what actually happens.

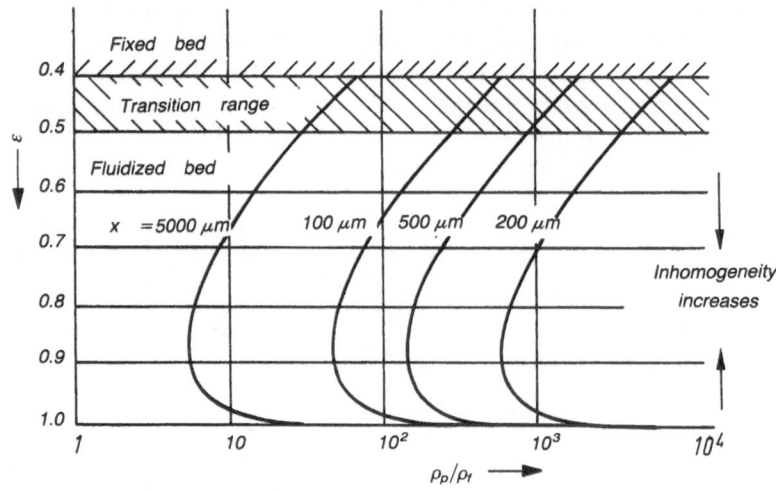

Figure 3.16 Stability diagram for a bed fluidized with air at 20 °C. The curves are valid for $S_{cr} = 0.14$. To the left of each S_{cr} curve stability (homogeneity) prevails for the given particle size ($S > S_{cr}$); to the right instability (inhomogeneity) prevails ($S < S_{cr}$).

3.4 THE CONTINUUM MECHANICS OF PACKINGS: POWDER MECHANICS

The principal ways in which packing behaves in response to the action of forces can be classified as on p. 55: as a compact body with the particles remaining in fixed positions, as a plastic body, and as a flowing medium.

In all three cases the packing is considered as a mechanical continuum and the stresses, deformations and rates of deformation arising are represented by the methods of elasticity theory, plasticity theory and fluid mechanics. The continuum approach has long been common practice in soil mechanics. Jenike [53] was the first to apply this approach to the study of bulk solids and thereby to deal theoretically with the flow of such solids in bunkers. This topic has been reviewed at length by Schwedes [25].

The problems posed in these two fields are of a quite different nature. In the case of flow in bunkers the stresses of interest are essentially lower than those occurring in soil mechanics. In soil mechanics we are interested in stable static behaviour. In mechanical process technology we are usually attempting to induce flow in, for example, mixers, chutes and bunkers. Moreover, the materials that have to be investigated are different. In soil mechanics the constituents of soil that are of interest are mainly coarse,

Section 3.4 was written with the cooperation of H. P. Kurz.

sands and fine-grained moist clays. In process technology we are concerned with very different types of bulk solids such as flour, cement, paint pigments, coarse plastic granules and the like, as well as moist materials such as filter cakes and agglomerates. Here a distinction must be made between dry bulk solids and soils.

3.4.1 Coefficient of pressure at rest

The coefficient of pressure at rest is a descriptive index used in soil mechanics. It shows that the plastic flow of packings cannot be regarded as analogous to the behaviour of liquids. Let us consider a semi-space of infinite extent with a horizontal surface in which the x axis lies. The y axis points vertically downwards. Let σ_{xx} be the horizontal and σ_{yy} the vertical compressive stress at any given depth. For reasons of symmetry these stresses are also principal stresses. The coefficient of pressure at rest is $\lambda_0 = \sigma_{xx}/\sigma_{yy}$.

In a liquid the molecules are free to move in any way. Therefore when the liquid is at rest the same hydrostatic pressure prevails on all sides, i.e. $\lambda_0 = 1$. In a purely elastic solid there exists under the action of its own weight a load in the vertical y direction whereas no load is applied in the x direction; therefore $\lambda_0 = 0$ for solids. A packing behaves neither like a liquid nor like a compact solid. Under the action of gravity the particles are also displaced in the horizontal direction so that an additional horizontal pressure σ_{xx} arises. However, the particles do not have complete freedom of movement like the molecules of a liquid, and so it follows that $\sigma_{xx} < \sigma_{yy}$ and $0 < \lambda_0 < 1$. A typical value is $\lambda_0 = 0.5$. The limiting values hold for the extreme conditions of the packing. For the fluidized bed $\lambda_0 \to 1$ and for highly compacted pressings $\lambda_0 \to 0$.

3.4.2 State of stress

If a continuum is subjected to static or dynamic forces, stresses are set up in it. In any given sectional area of a continuum there exist normal and tangential stresses which, integrated over the area, must be in equilibrium with the forces being applied. If an element of volume within the continuum is considered, then there act on it, on the one hand, forces proportional to the mass of the volume, such as the gravitational force and the inertial force, and, on the other hand, surface forces due to the stresses. If the volume of the element approaches zero the forces due to its volume vanish. The surface forces alone then remain in equilibrium with one another and represent the state of stress in the body at the point being considered. The state of stress is independent of the choice of coordinates or boundary surfaces for the element of volume.

The components of the state of stress (stress tensor), i.e. the **normal stresses** and **tangential stresses** acting on any given surface element, depend, of course, on the orientation of this element. For the complete designation of the three-dimensional state of stress three normal stresses and three equal pairs of tangential stresses are sufficient; these act on the surfaces of a cuboidal element of volume oriented in any direction, i.e. there are in all six components of stress. For a particular orientation of the volume element the tangential stresses vanish. The state of stress is then completely specified by the three normal stresses σ_1, σ_2 and σ_3 called the **principal stresses**. They act in the principal directions which are perpendicular to the surfaces of the volume element. Therefore in order to specify completely the state of stress at a point in a three-dimensional field of stress produced by any forces six quantities are necessary: the three principal stresses and three angles specifying the principal directions.

In the theory of elasticity and plasticity, as well as in fluid mechanics, tensile stresses are chosen to be positive and compressive stresses are chosen to be negative. In soil and powder mechanics the compressive stresses predominate; therefore it is customary to define compressive stresses as positive. This should be kept in mind.

If the forces occurring in a packing are to be described by the methods of continuum mechanics then the volume element must be large compared with the particle size, but generally comparable with the size of any existing agglomerates. Another requirement is that the extent of the part of the packing being considered, e.g. the size of the equipment, should be large in comparison with the volume element.

The manipulation of three-dimensional states of stress, which is more difficult, will not be considered here. We shall restrict ourselves to two-dimensional states of stress. To describe these we shall choose a volume element with a right-angled triangular cross-section whose shorter sides are perpendicular to the direction of the principal stresses and whose hypotenuse is perpendicular to the x direction in the one instance and to the y direction in the other. We shall adopt the convention that the larger of the two principal stresses is designated σ_1; hence σ_1 will always be greater than σ_2. The normal stresses in the x and y directions are σ_x and σ_y; both tangential stresses τ_{xy} have the same value.

The following equations are obtained by considering the equilibrium of the forces:

$$\sigma_x = \sigma_1 \cos^2 \alpha + \sigma_2 \sin^2 \alpha = \tfrac{1}{2}\{(\sigma_1 + \sigma_2) + (\sigma_1 - \sigma_2)\cos 2\alpha\} \quad (3.59a)$$

$$\sigma_y = \sigma_1 \sin^2 \alpha + \sigma_2 \cos^2 \alpha = \tfrac{1}{2}\{(\sigma_1 + \sigma_2) - (\sigma_1 - \sigma_2)\cos 2\alpha\} \quad (3.59b)$$

$$\tau_{xy} = \tfrac{1}{2}(\sigma_1 - \sigma_2)\sin 2\alpha \quad (3.59c)$$

The equations can be represented by the Mohr stress circle (Fig. 3.17). The shear stress τ_{xy} is plotted on the ordinate and the normal stresses σ_x and σ_y are plotted on the abscissa. The centre of the Mohr stress circle is at

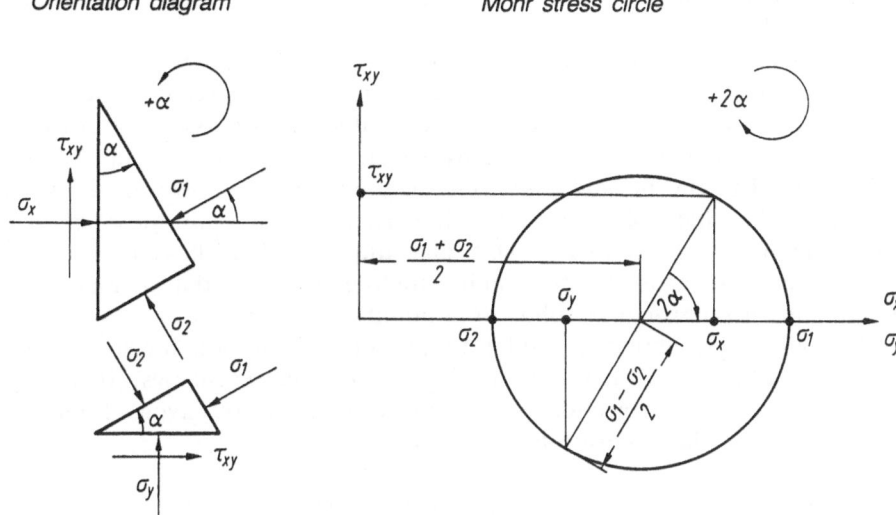

Figure 3.17 Representation of a two-dimensional state of stress by the Mohr stress circle.

$(\sigma_1 + \sigma_2)/2$ on the abscissa and its radius is $(\sigma_1 - \sigma_2)/2$; hence the points of intersection with the abscissa give the principal stresses σ_1 and σ_2. The normal stress σ_x and the associated shear stress τ_{xy} are fixed by the radius arm which is rotated through an angle 2α from the radius arm of the larger principal stress σ_1, i.e. from the abscissa. It should be noted that in the Mohr stress circle the sense of rotation of the angle 2α is positive in the clockwise direction, as opposed to that shown in the orientation diagram. The normal stress σ_x now appears as the projection of the radius arm onto the abscissa and the shear stress τ_{xy} appears as the projection onto the ordinate. σ_x and σ_y include the angle 90°. Since the double angle occurs in the Mohr circle, the radius arms ascribed to the normal stresses form a diameter. The normal stress σ_y also appears as a projection on the abscissa. It can be seen from the Mohr circle that the shear stresses have their maximum values for $\alpha = 45°$ and $\alpha = 135°$.

3.4.3 Yield criteria for packings

If a solid body is compressed by a normal force N against a flat support and subjected to a tangential force S, then it will remain at rest as long as $S < \mu N$ and will move if S exceeds a limiting value; it begins to move when $S = \mu N$ ($\mu = \tan\delta$ is the coefficient of friction and δ is the angle of solid friction). Strictly speaking, a distinction should be made between static friction and dynamic friction, i.e. the friction when movement occurs. In this case the frictional force cannot become greater than μN. This is the Coulomb friction criterion for solid bodies. It is a matter of practical experience that it is fulfilled over a wide range of sliding velocities.

In a manner analogous to that indicated when considering the friction between solid bodies, the yield or flow criteria for packings show how those shear stresses that lead to irreversible, i.e. plastic, deformation depend on the normal stress. This shear stress at the yield point is proportional to the normal stress only in special cases; however, for small velocities the shear stresses at which particulate materials flow are also independent of the velocity. This and the fact that the coefficient of pressure at rest λ_0 is less than unity are the two essential characteristics that distinguish particulate solids from liquids.

A yield criterion that is similar to Coulomb's criterion for the friction between solid bodies has long been used in soil mechanics. It will be discussed in the next section. Later, an account will be given of Jenike's modification to this criterion for bulk solids.

(a) The Coulomb yield criterion used in soil mechanics

The general form of the expression for the Coulomb yield criterion used in soil mechanics is

$$\tau = \sigma \tan \delta + c \qquad (3.60)$$

where τ is the shear stress at which the soil begins to deform plastically under a normal stress σ, δ is the angle of friction and c is the 'cohesion', which is a measure of the stickiness of the solids, i.e. a measure of the ratio of the adhesive forces between the particles to their weight. The designation 'cohesion' is misleading since it does not represent a tensile stress but a shear stress when the normal stress is zero. Despite this the generally accepted concept will be retained here and also later when dealing with the yield criteria for bulk solids.

The graph of the Coulomb yield criterion on the (σ, τ) diagram is a straight line (see Fig. 3.18). Its intercept on the ordinate is the cohesion. For cohesionless solids such as dry sand the cohesion disappears and we obtain the special case of a Coulomb line going through the origin and having only one characteristic parameter, i.e. the angle of friction δ.

This angle of friction can be demonstrated experimentally. If a container that has been filled to the rim with a cohesionless material is gradually tipped over, the granules begin to trickle over a sloping surface as soon as a certain angle of inclination is reached (Fig. 3.19). If the tilt of the vessel is now increased, the angle of inclination of the granular material, i.e. the angle of repose γ, is maintained. The angle of repose is equal to the angle of friction for cohesionless solids. The gravitational component of force exerted by the uppermost layer of particles on the sloping surface is very

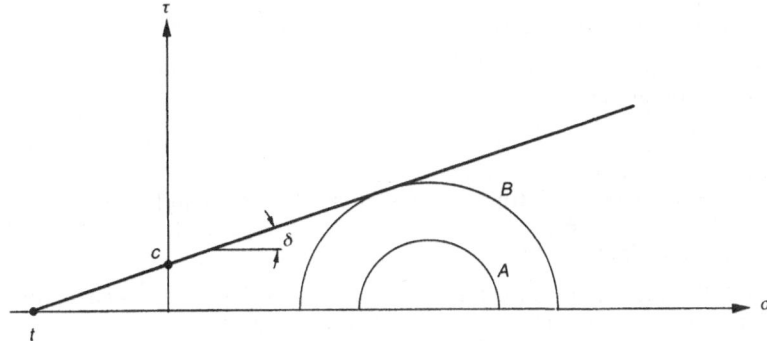

Figure 3.18 Representation of the yield criterion according to Mohr and Coulomb (see the text).

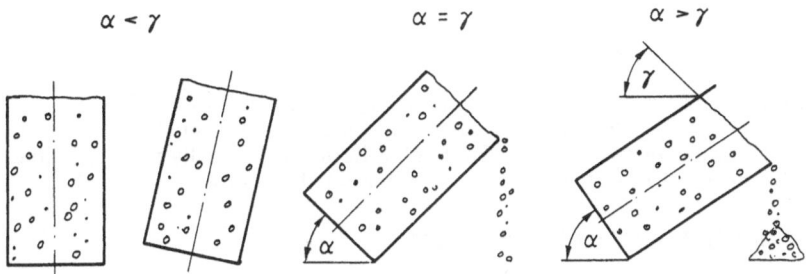

Figure 3.19 Sketch of the internal angle of friction of cohesionless bulk solids.

small. The tangential shear stress is correspondingly also small. Despite this the layer slips. It is a matter of experience that for cohesive solids the angle of repose is less reproducible and is not a measure of the angle of friction. This can be shown to be plausible by means of a similar experiment. A cohesive material, e.g. moist sand, is tamped into a container which is then tilted through 90° without any of the material flowing out. However, an angle of friction of 90° is not possible because otherwise in a shear test the shear stress would have to become infinitely large. In fact, the material is retained by adhesive forces; the internal normal stress in the slip plane is not zero.

The Coulomb line thus denotes those related τ–σ stresses at which yield occurs. Above the Coulomb line no states of stress are possible because at a given σ the material begins to flow as soon as the matching τ of the Coulomb line is reached.

Each state of stress can be represented on the (σ, τ) diagram by a Mohr stress circle (Fig. 3.18). Thus the following conditions apply:

1. States of stress with a Mohr circle A below the Coulomb line cause no plastic deformation;
2. States of stress with a Mohr circle B tangential to the Coulomb line cause plastic deformation since this Mohr circle includes precisely that pair of (σ, τ) values necessary for plastic deformation to occur;
3. States of stress with Mohr circles that cut the Coulomb line are physically impossible since such Mohr circles would have pairs of (σ, τ) values above the Coulomb line.

The following relations for the principal stresses that lead to plastic flow (yielding) are obtained by simple geometrical analysis:

$$\sigma_1 = \sigma_2 \frac{1 + \sin \delta}{1 - \sin \delta} + 2c \frac{\cos \delta}{1 - \sin \delta} \tag{3.61a}$$

$$\sigma_1 = \sigma_2 \tan^2 (\delta/2 + 45°) + 2c \tan (\delta/2 + 45°) \tag{3.61b}$$

For the special case of cohesionless soils with $c = 0$ the relation is reduced to

$$\frac{\sigma_1}{\sigma_2} = \frac{1 + \sin \delta}{1 - \sin \delta} \tag{3.61c}$$

In soil mechanics the pressures lie in a range up to and over 20 bar (2×10^6 N/m^2), whereas in powder mechanics they are usually below 1 bar. For this range of pressure the Coulomb criterion is generally not applicable. Only cohesionless soils also exhibit Coulomb behaviour at low pressures.

(b) Yield behaviour of bulk solids

The yield criteria for bulk solids were introduced by Jenike. The yield behaviour depends upon the existing state of consolidation of the bulk solid (particulate material) when it is caused to flow or yield under the action of normal and shear stresses. The various kinds of yield behaviour can be observed and measured using the Jenike shear test. Jenike's shear cell is represented schematically in Fig. 3.20(a). It consists of a lower fixed ring and an upper movable ring. The material to be investigated occupies the common cylindrical space within the rings and has a particular porosity ε that has been determined by the prior filling process or adjusted by an additional compaction. The material is subjected to a constant normal force N applied through a lid. A shear force S is applied by means of a bracket attached to the lid and is registered on an $x–t$ recorder. It is possible to register three fundamentally different shear force–displacement curves (Fig. 3.21).

1. Initially the material deforms elastically and the shear force increases rapidly (a). Beyond a certain value it remains constant while the material deforms plastically, i.e. yields by flowing. During this yielding

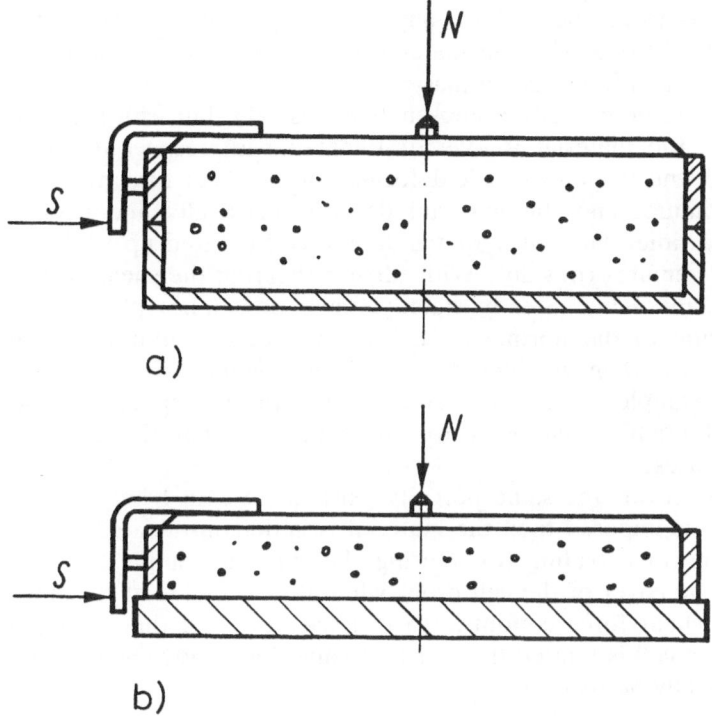

Figure 3.20 Sketch showing the principle of the Jenike shear tester: (a) material–material friction; (b) material–wall friction.

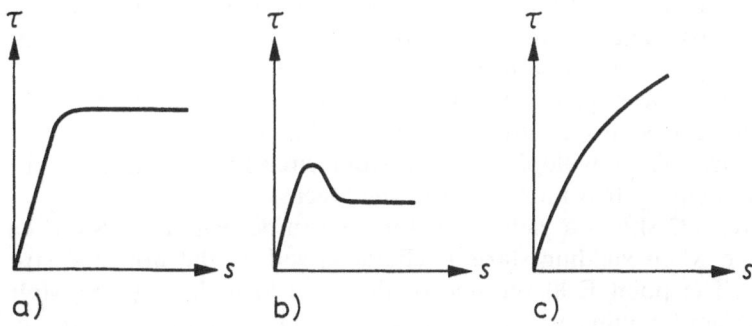

Figure 3.21 The three possible shear stress versus displacement plots.

porosity remains constant. Such behaviour is only possible when a particular normal force is applied that is specific to the porosity in question. The material is then described as **critically** consolidated with respect to this normal loading. This shearing process simulates the case of steady state yielding under which the porosity is constant, i.e. time independent, at a given place;

2. If the same porosity is chosen as in case (1), but with a smaller normal load, the material is overconsolidated with respect to this load (b). After the range of elastic deformation the shear force passes through a maximum, and the material deforms plastically and expands at the same time. The cover of the Jenike cell is lifted up while the normal load remains constant. With further shearing the shear stress falls to a constant value. Steady state yielding now takes place at a porosity specific to the normal load. Such a shearing process describes what happens when yielding starts and the volume expands as is the case, for example, after a bunker is opened and the upper level of the bulk solid remains stationary for some time so that the average porosity increases;

3. If, however, the same porosity exists as in case 1 but a higher normal force is imposed then the material is underconsolidated with respect to the normal loading (c). During shearing the sample becomes denser, i.e. the cover of the cell drops while the normal load remains constant. The shear force continues to increase. Because the shear path of the Jenike cell is limited the constant value for steady state yielding cannot generally be reached.

With many problems in process technology we are interested not only in the behaviour during steady flow but also in the behaviour when yielding starts. In the Jenike test several pairs of τ–σ values, which appertain to the states of stress when yielding starts, can be determined at constant porosity under various normal loads. Here τ corresponds to the maximum value in Fig. 3.21(b). The τ_E–σ_E values of steady state flow can be measured similarly. τ_E is the constant shear stress in Fig. 3.21(a). As Schwedes [25] has shown, these pairs of values can be assigned to Mohr circles that represent the states of stress when yielding begins or steady state flow takes place. The envelope of these Mohr circles is generally not a straight line but a curve. It is known as the **yield locus**.

Figure 3.22 shows a yield locus for a given porosity ε. A Mohr circle for the stage when yielding starts is characterized by the principal stresses $\bar{\sigma}_1$ and $\bar{\sigma}_2$. The point E at the end of the yield locus lies on the Mohr circle (σ_1, σ_2) appertaining to steady state flow. The point c gives the cohesion of the material at the given porosity and the point t the tensile strength. There is also a Mohr circle that starts at the origin and touches the yield locus. Its second point of intersection with the abscissa is designated f_c. The corresponding state of stress has obviously only the one principal stress f_c other than zero. This circle represents a uniaxial compression test; f_c is thus the compressive strength of the material.

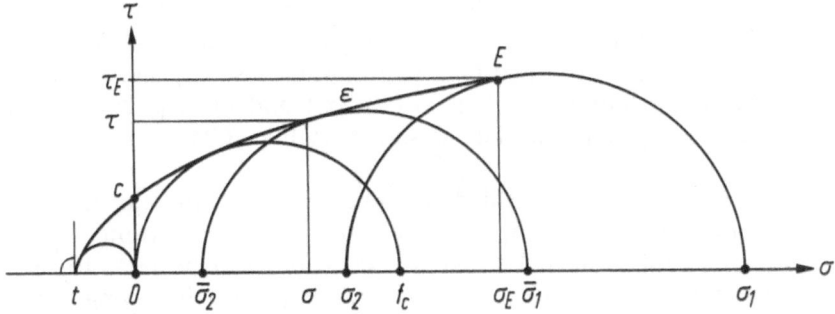

Figure 3.22 Jenike's yield locus.

For each porosity there is a particular yield locus; a family of three yield loci is shown in Fig. 3.23. It has been established by many experiments that the envelope of the Mohr circles through the points E_i that lead to steady state flow is, to a very close approximation, a straight line through the origin. This line thereby represents an extremely simple and convenient physical law for steady state flow. It is called the **effective yield locus**. Its angle of inclination δ_e is known as the effective angle of friction. Therefore, corresponding to eqn (3.61c), we have the relation

$$\frac{\sigma_1}{\sigma_2} = \frac{1 + \sin \delta_e}{1 - \sin \delta_e} \tag{3.61d}$$

which can now be regarded as the physical law for steady state flow.

In addition to this relation, the factors of interest concerning the problems of process technology are the porosity ε prevailing during steady state flow and the proper compressive strength f_c as functions of the

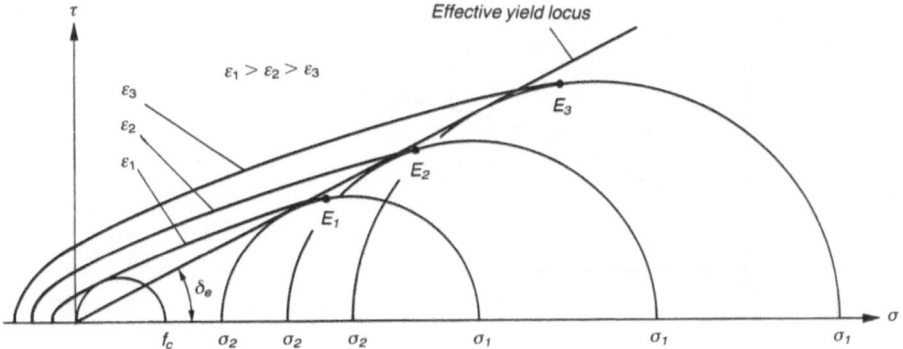

Figure 3.23 Family of yield loci with effective yield locus.

consolidating pressure. For this it is expedient to choose the larger of the principal stresses σ_1 of steady flow for the yield laws appertaining to ε and f_c. Each yield locus is specified by a pair of values (ε, σ_1) and another pair (f_c, σ_1). From the family of yield loci (ε, σ_1) and (f_c, σ_1) curves can be constructed for a given material, as shown in Fig. 3.24. A material flows poorly if its strength f_c is large relative to the consolidating stress σ_1 or the ratio σ_1/f_c is small. Jenike called this ratio the flow function (FF). It is generally not a constant since, as a rule, the compressive strength f_c is not proportional to the principal stress σ_1. The ratio can be used to make a rough classification of bulk solids according to the following different ranges: FF < 2 for cohesive non-flowing materials; 2 < FF < 4 for cohesive materials; 4 < FF < 10 for ready flowing materials; 10 < FF < ∞ for freely flowing materials.

The angle of wall friction ϕ' for a bulk solid can also be measured in a shear cell (Fig. 3.20(b)). Experimental results have shown that, to a close approximation, the wall shear stress τ_W is proportional to the normal stress σ_W, i.e. for a given material Coulomb's relation

$$\frac{\tau_W}{\sigma_W} = \tan \phi' \tag{3.62}$$

is valid.

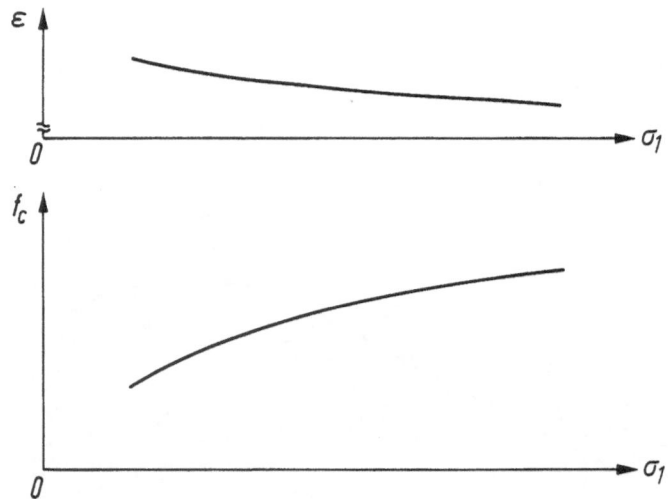

Figure 3.24 The dependence of the porosity ε and the compressive strength f_c on the largest principal stress σ_1 occurring during steady state flow.

3.4.4 The design of bunkers using Jenike's method

Two different kinds of flow, i.e. core flow and mass flow, can occur in bunkers, as shown in Fig. 3.25. In the first case the material flows within a restricted zone. Core flow occurs in relatively flat hoppers. The disadvantages of this are the tendency for the material to segregate, for the flow to be irregular and for material to remain in the peripheral region for a long time. With mass flow, which occurs in steeply sloping hoppers, all these disadvantages disappear. Jenike has developed a method for (a) determining the slope required in the hopper to ensure mass flow and (b) calculating a lower limit for the aperture of the discharge. This limit is that size at which the formation of stable bridges of material starts to occur. Within the scope of the present text the theoretical arguments and extensive results can only be outlined.

According to Jenike, the limiting hopper slope θ'_{max}, measured from the vertical, that just allows mass flow can be read off plots of the function

$$\theta'_{max} = f(\delta_e, \phi') \tag{3.63}$$

Mass flow prevails for $\theta' < \theta'_{max}$.

In order to establish a criterion for the formation of a stable bridge of material in the outlet it is assumed that such a bridge exists and is loaded by an applied stress σ'_1. The bridge supports itself. The second principal stress in the bridge is then zero. The applied stress σ'_1 is calculated from the expression

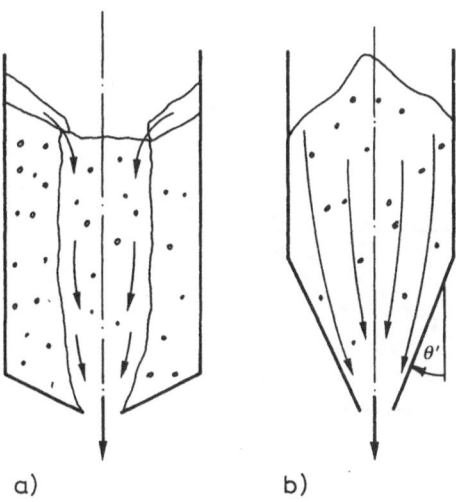

a)   b)

Figure 3.25 Sketch illustrating (a) core flow and (b) mass flow.

$$\sigma_1' = \frac{(1 - \varepsilon)\rho g d}{H(\theta')} \tag{3.64}$$

where d is the outlet diameter of a rotationally symmetric bunker or the slot width of a flat-sided bunker, ρ is the density of the solid, g is the acceleration due to gravity and $H(\theta')$ is a function specified by Jenike for both bunker geometries. This applied stress subjects the bridge to pressure. If it is larger than the compressive strength f_c the bridge will collapse. Furthermore, it can be shown that, near the outlet, the ratio of the larger principal stress σ_1 under steady state flow to the stress σ_1' acting on the bridge is constant for a given bunker geometry; the relation

$$\text{ff} = \frac{\sigma_1}{\sigma_1'} = \text{ff}(\theta', \delta_e, \phi') \tag{3.65}$$

holds.

Jenike has specified how the **flow factor** ff depends on the variables θ', δ and ϕ'. The applied stress $\sigma_1' = (1/\text{ff})\sigma_1$ is determined from the flow factor ff, and the compressive strength $f_c = (1/\text{FF})\sigma_1$ is determined from the flow function FF. When σ_1' and f_c are plotted against σ_1 (Fig. 3.26), the two curves intersect at a single point. In accordance with Jenike's criterion for stability this point of intersection separates the region where bridges form ($\sigma_1' < f_c$) from that where they do not form ($\sigma_1' \geqslant f_c$). The critical outlet dimension d_{crit} can be obtained from the appropriate critical applied stress $\sigma_{1,\text{crit}}'$ by the relation

$$d_{\text{crit}} = \sigma_{1,\text{crit}}' \frac{H(\theta')}{(1 - \varepsilon)\rho g}$$

By choosing $d \geqslant d_{\text{crit}}$ the formation of bridges can be avoided.

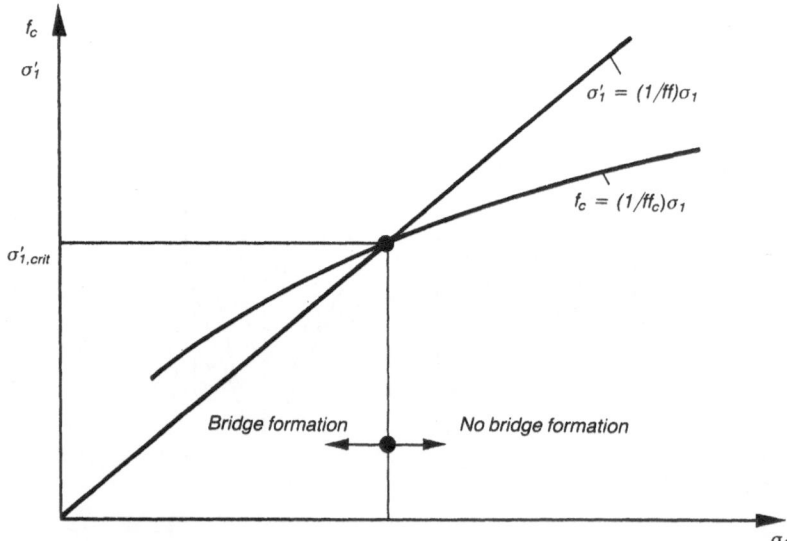

Figure 3.26 Diagram for determining the critical outlet dimension to prevent the formation of bridges.

3.5 DEFORMATION AND FRACTURE OF SOLIDS

Disperse solid phases are generally produced by the comminution of coarse feed material. The processes of fracture that destroy a particle can be actuated only by elastic fields of stress within it. External forces give rise to a three-dimensional state of stress which is determined by the shape of the particle and the distribution of the load on it, and which is relieved either by the yield point being exceeded in places and causing an irreversible plastic deformation of the particle or by compliance with the conditions that lead to fracture. Frequently the condition for plastic flow is first satisfied, at any rate at those points where the external forces act. Multidimensional yield processes produce, on their part, stress fields in their surroundings which, in the case of materials with a large interval between the limits of yield and fracture, can influence the course of breakage decisively. More will be said about this later. Fractures begin at structural flaws; the larger these are, the more they endanger the material. With decreasing particle size the probability of large cracks being found diminishes; as a consequence there is a decreasing chance that fracture occurs. However, the yield point is less strongly dependent on the particle size. These size-dependent attributes result in an increasing irreversible deformability with decreasing particle size; for example, limestone particles smaller than $10\,\mu$m and quartz particles smaller than $3\,\mu$m undergo significant plastic deformation before breaking [84].

3.5.1 Fracture criteria

Real solid bodies do not possess a perfect structure; under loading, stress concentrations which can initiate fracture arise at structural flaws. The fracture process is a matter of energetics. Energy is needed to generate new boundary surfaces. This energy comes from the field under stress and the thermal energy of the body. Further energy-donating mechanisms can be adsorption and/or chemical reactions on the newly created fracture surfaces. In the case of stress corrosion additional contributions come from the chemical potential.

The consumption of energy is due mainly to irreversible deformation in the immediate neighbourhood of the crack tip. This is true even for brittle materials such as glass. The energy consumption related to the fractured area is defined as the specific fractured-area energy β whose values range from 5×10^{-4} to 10^{-3} J/cm^2 for glasses, from 10^{-3} to 10^{-1} J/cm^2 for plastics and from 10^{-2} to 10 J/cm^2 for metals. The specific surface energy γ of solids lies in the range 10^{-6}–10^{-4} J/cm^2; hence it follows that γ is always very much less than β. Other energy-consuming processes are structural alterations and the occurrence of electrical disturbances which may lead to

Section 3.5 was written with the cooperation of Professor K. Schönert.

the emission of light and elementary particles. A rapidly advancing crack is a source of elastic waves whose energy is often simply considered to be the kinetic energy of the fracture; they are to be included among the consumers of energy. The energetics involved have been discussed at length by Rumpf [78].

Griffith was the first to set up an energy balance in that he put the available strain energy for a differential advance of a crack in equilibrium with the necessary surface energy. Since the former is proportional to the length of the crack, while the specific energy consumption remains constant, a characteristic instability arises. Beyond a critical crack length there is an excess of energy and the crack can suddenly extend without being arrested, a phenomenon that is indeed observed experimentally. Fracture mechanics developed during recent decades has concerned itself with the problem of the energy available in the stress field [73]. We shall relate this energy to the cross-section of the fracture (the fracture cross-section is equal to half the surface area of the crack) and call it the specific fracture-propagating energy G. For a fracture in a flat tensile-loaded sample of material of unit thickness with a crack length l (Fig. 3.27)

$$-\frac{\partial U_{el}}{\partial l} = G = \pi(1 - v^2)\sigma^2\frac{l}{2E} \tag{3.66}$$

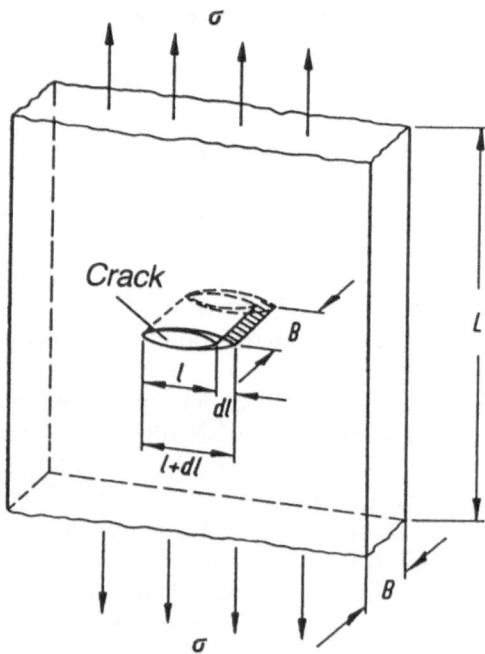

Figure 3.27 Sketch of a tensile test sample with a crack.

where U_{el} is the elastic energy per unit thickness of sample, v is Poisson's ratio, σ is the unaffected tensile stress and E is Young's modulus. The fracture starts when the condition $G \geqslant 2\beta$ is satisfied.

It can be seen from eqn (3.66) that the available energy, represented by G, increases in proportion to the crack length only when the nominal tension σ remains constant. However, the crack consumes energy; σ is therefore reduced and the external forces may provide fresh energy. However, this cannot occur if the fracture grows rapidly. The size of the sample determines the decrease in the stress during the propagation of the fracture since the energy reserve is proportional to the volume while the consumption of energy is proportional to the cross-section. The smaller is the sample, i.e. particle, the smaller is the ratio of stored energy to consumed energy and the greater is the reduction in σ. This can result in the fracture coming to a standstill unless more energy is supplied from outside.

These relations can easily be assessed when a tensile test sample whose length L is varied is fractured. An estimate of the condition that must be satisfied in order that the energy stored before the fracture starts can force it right through the sample is that L must be of the order of $4\beta E/\sigma_z{}^2$. The minimum lengths for steel ($\sigma_z = 5 \times 10^4$ N/cm^2, $\beta = 10$ J/cm^2) and glass ($\sigma_z = 10^4$ N/cm^2, $\beta = 10^{-3}$ J/cm^2) are found to be 32 cm and 0.28 mm respectively. In investigating the strength of materials we generally use samples that are so long that this size effect can be neglected; however, it exerts a greater influence when small grains are being crushed.

In the immediate neighbourhood of the tip of a fast-advancing crack large plastic deformations occur in fractions of microseconds. This leads to considerable local heating of the material. Calculations and experiments have shown that this can result in temperature rises of about 1000 °C or more. The local behaviour of the material differs completely from that of the surroundings. The residual state deviates considerably from that of a surface in equilibrium; it possesses a different structure and morphology. Comparison of adsorption measurements, before and after contact with air, on fracture surfaces produced under vacuum gives an indication of the changes produced in the surface by the action of water vapour. The amount of krypton adsorbed on glass, quartz, calcite and sugar is reduced by factors of 2.7, 1.5, 1.6 and 3.4 respectively [26].

3.5.2 Deformation and fracture of particles under stress

In a pulverizer particles of irregular shape are stressed by pressure, friction and impact against solid bodies or by gliding movements such as shear flow. In general the particles undergo elastic and/or plastic deformation and the resultant stress field depends on the nature of the loading, the deformation behaviour, the shape of the particles and their structural inhomogeneities. Different fracture phenomena and differing size distributions of daughter particles follow as a result of the varying fields of stress. This is the reason why no generally valid function that describes the size distribution of daughter particles and the generation of new surfaces can be found. The basic phenomena have been studied using spherical particles [20].

In the extreme case of purely elastic deformation, curved fracture trajectories are observed that run from one contact zone to that lying directly opposite (Fig. 3.28) or, under impact loading, propagate like rays from the cluster of fractures at the point of impact. The fractures approximately follow those principal stress trajectories that are perpendicular to the maximum tensile stresses. This primary fracture phase is followed by a secondary stage that, even in the case of a static compressive stress, is determined by dynamic stress fields which are produced by the rapidly advancing fractures. If primary fractures relieve the stress in the compression zone directly below the contact surfaces sufficiently rapidly, vibrations that result in tensile stresses and perhaps secondary fractures are induced in the material. As the theory shows, these regions possess the highest energy density before fracture begins; when this energy is released it results in the production of mainly fine material. In the case of impact

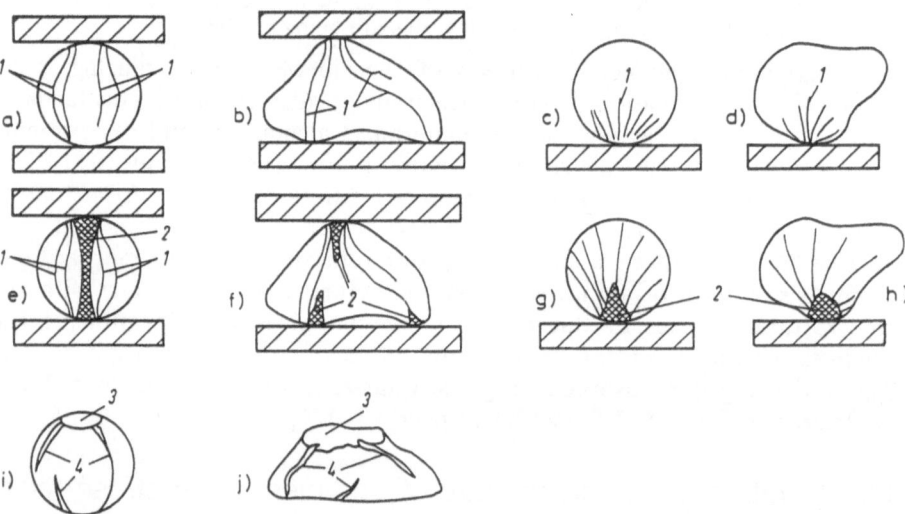

Figure 3.28 Breakage phenomena induced by compression and impact: (a)–(j) fractures occurring under various stressings of the particles; 1, primary fractures associated with elastic deformation; 2, zone of fine material; 3, contact surfaces formed by plastic deformation; 4, meridional fractures associated with plastic deformation.

loading a cone producing fine material is also observed in this region of high energy density below the centre of impact. Fractures occurring at wavefronts or where reflected waves converge are to be expected only with impact velocities of several hundreds of metres per second; in other cases the stress intensities are not high enough to produce fractures during the short transit times of the waves.

If a material deforms mainly plastically then a cone-shaped zone is pushed into the centre of the sample and the material on the side of it is forced outwards. These movements of material give rise to peripheral tensile stresses perpendicular to the directions of movement and, in the case of spheres, perpendicular to the meridional planes with the consequence that the fractures break the sphere up into pieces like the segments of an orange. The penetrating cone-shaped pieces can often still be recognized even after the particle has been destroyed. The domain directly under the point of contact is either only slightly shattered or is not shattered at all; in contrast with the case discussed above of preponderantly elastic behaviour no fine material is formed. This mechanism can also be observed with impact loading if the material is still capable of deforming irreversibly when subjected to a load over a very short time.

In the case of plastic deformation, tensile stresses are frequently produced in the elastic compressive stress zone of high energy density and are superimposed on the compressive stresses during loading, with the result that no fracture occurs. When the pressure is released, the tensile stresses produced by the plastic deformation survive as internal stresses and produce **relieving fractures**.

The deformation behaviour is not merely a property of the material but is also influenced by the temperature, the rate of stressing and the sample size. The quicker is the loading and the lower is the temperature of the sample, the shorter is the plastic stage; the material turns brittle. However, decreasing the particle size increases the plasticity; even materials like sapphire, quartz and glass can be plastically deformed if the size of the stressed particle does not exceed a few micrometres [87]. In tiny particles the cracks are naturally smaller, and larger stresses must be developed to satisfy the differential energy condition. Similar behaviour is required by the integral energy condition (see section 3.5.1).

Stressing in a shear flow is very different from that just discussed. Raasch [19] has calculated the state of stress for isotropic spherical particles. It is a pure state of shear stressing with equally large maximum compressive, tensile and shear stresses that are 2.5 times as great as the shear stress in the shearing stream (Fig. 3.29). The particle rotates with an angular velocity ω that is equal to half the rate of shear κ, i.e. $\omega = \kappa/2$. The particle is thus subjected to an alternating stress whose level, however, is generally not sufficient to destroy compact grains but which can break up agglomerates. Agglomerates are then abraded layer by layer from the

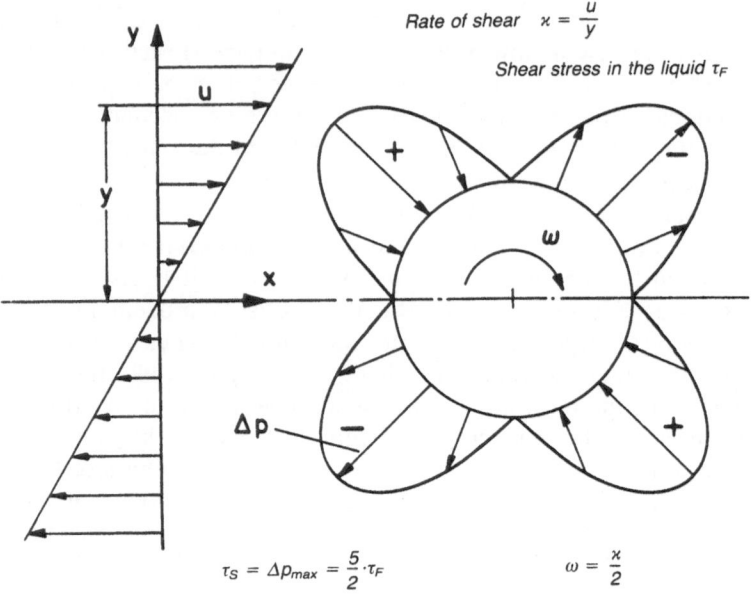

Figure 3.29 Pressure distribution for a sphere in a shear flow [19].

outside; they shrink and a trail of detached primary particles arises behind them. The breaking up of agglomerates depends not only on the shear stress but also mainly on the frequency with which local compressive and shear stresses change as a consequence of the rotation of the grains. The latter effect is taken into account by the stress cycle number, which is defined as the product of the frequency (cycles per second) and the duration of the stress in a crevice [58, 77].

3.5.3 Comminutive properties of materials

To design and evaluate a crushing process it is necessary to know the fracture probability of the grains and their resistance to fracture, the size distribution of the daughter particles, the increases in surface area due to comminution and the state of agglomeration after crushing. We call these factors the comminutive properties of matter; they depend on the material, the nature and intensity of stressing, the rate of stressing, the temperature, the surroundings, the type of stressing device used and its shape, and the particle size [79]. The investigation of this very complex matter is far from complete. Experiments on single grains have so far produced the following findings.

The probability that a material will fracture depends upon the size of the particles and the intensity with which they are stressed. The intensity is expressed as the energy expended in pulverizing a unit volume of material: $E/V = E_v$. For any given value of E_v the probability of fracture decreases as the particle size becomes smaller. Typical results for the impact stressing of glass spheres on flat plates are shown in Fig. 3.30. These results are well represented by straight lines plotted on log-probability graph paper. By analogy with the above, the specific reactive force under compression increases as the particle size decreases; Fig. 3.31 shows the results for quartz particles in the size range from 10 μm to 10 mm. The parameter introduced here is the reduction ratio which is defined as the ratio of the particle size to the final gap between the compression plates. The reduction ratio is an appropriate index for the practical problems of crushing because in many pulverizers the final gap between the crushing implements is known in advance.

The results in both figures show how difficult it is to crush fine grains. In the case of impact stressing the energy per unit volume is given by $E_v = \rho v^2/2$, where ρ is the density of the material. With fine particles the velocity of impact v must be raised. The increasing influence of aerodynamic factors sets a limit on this endeavour. The compressive crushing of fine grains takes place in a bed of particles; as the size of the particles decreases this bed sustains larger and larger forces which thus have to be satisfied by the design of the equipment.

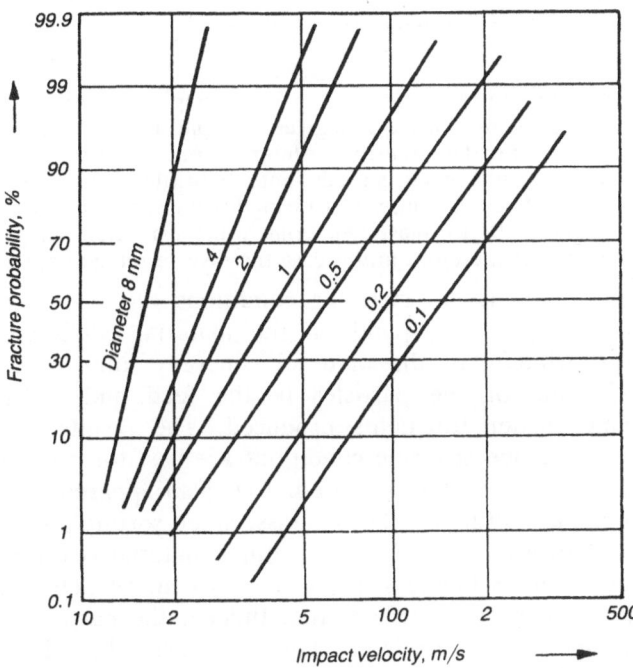

Figure 3.30 The probability of glass spheres breaking under impact (the parameter is the diameter of the spheres).

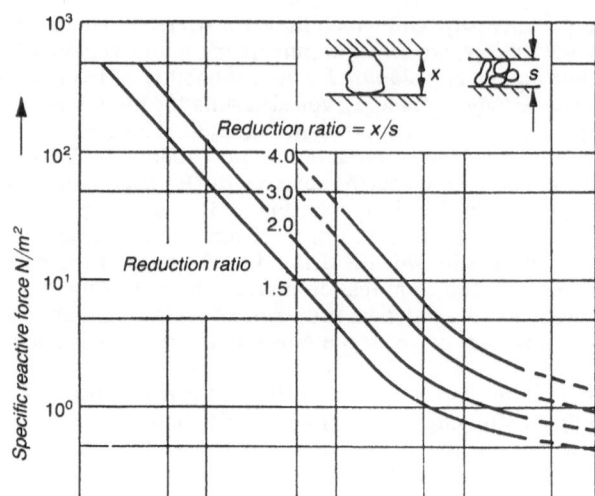

Figure 3.31 Specific reactive force of quartz grains subjected singly to compressive stressing (the parameter is the reduction ratio).

The size distribution of the daughter particles depends on the nature and intensity of the stressing. Figure 3.32 gives examples for 1.1 mm grains of limestone. The greater the intensity is, the more fines are produced. With this material, whose behaviour is predominantly brittle, compressive stressing gives rise to more fines than impact stressing for the same energy supply. This finding is not valid for plastic materials which can often only be shattered by impact; therefore in this case the inverse relation applies.

By appropriate choice of stressing conditions the size grading can be varied within limits that depend on the material being crushed. In a pulverizer the grinding of the material generally follows as a result of successive stressing of the particles in the feed and of the primary, secondary and further fragments produced. The results of single-grain crushing indicate which stressing conditions are best for obtaining a milled product with the size grading required. It is also apparent that not every requirement can be fulfilled without an associated sorting process.

The specific surface S_v or S_m of a ground material is often used as an integral measure of its fineness, particularly whenever subsequent processes are determined by the surface area. Indeed, the particle size distribution and the specific surface are related to each other. Because of the irregular shape of the particles, the difficulty of describing the roughness of their structures and the various definitions of surface area, the specific surface can only be calculated on an arbitrary basis, however. An

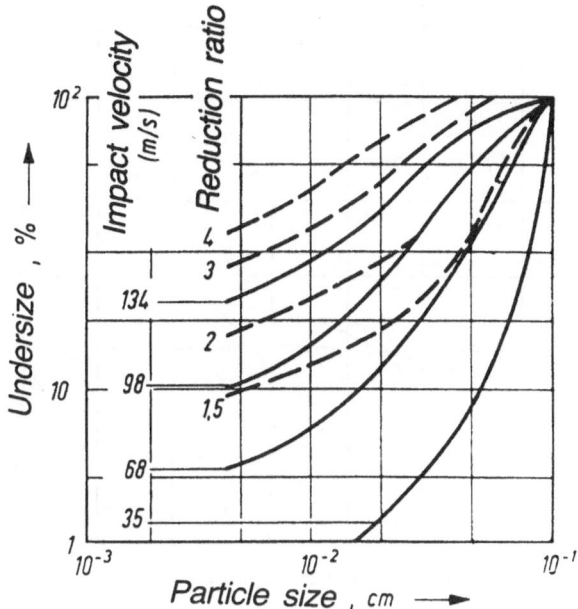

Figure 3.32 Particle size distribution produced by compressive (---) or impact (—) stressing of single 0.11 cm grains of limestone (the parameters are the reduction ratio of the velocity of impact).

additional complication arises from the fact that the finest fractions, whose distributions are difficult to measure, often contribute appreciably to the surface area. The increase $\Delta S/V$ or $\Delta S/M$ in the specific surface for defined stressing conditions and particle sizes is therefore treated as a separate comminutive property of matter. With mass products such as cement the problem is one of achieving a set specific surface by grinding with the least possible consumption of energy.

Figure 3.33 shows the utilization of energy, i.e. $\Delta S/E = S_E$, as a function of the mass-related energy E_M for both compressive and impact crushing of limestone. Crushing by compression utilizes energy better than crushing by impact. In both cases finer grains yield better values than coarse ones. With compressive crushing the utilization of energy decreases monotonically as the supply of energy increases because more and more work has to be done on account of the friction within the heap of fragments, and this process contributes less effectively to the comminution. Impact crushing, however, shows a maximum that is associated with a fracture probability of just under 100% and it should also be pointed out here once again that impact stressing is more effective than compressive stressing for those materials that deform mainly plastically.

Fine fragments are inclined to agglomerate. This is particularly true for a

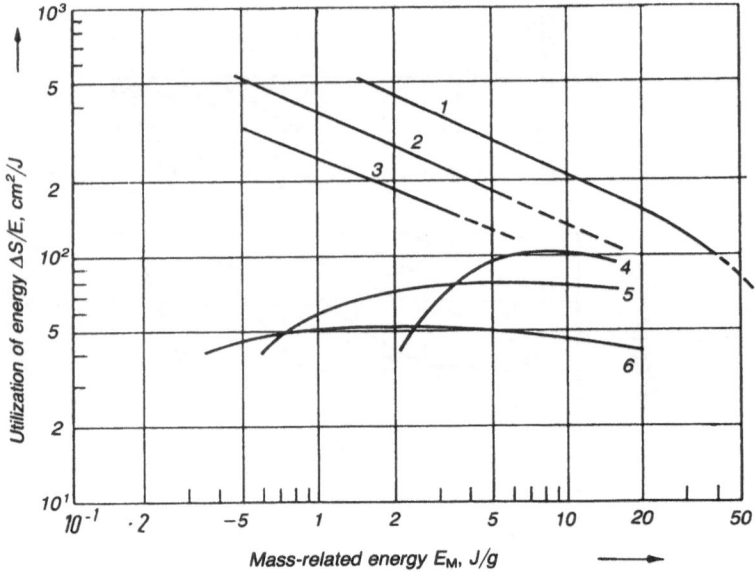

Figure 3.33 The utilization of energy in the stressing of single particles of lime-stone: 1, 2, 3, compressive stressing (1, particle size 0.1 mm; 2, particle size 1.0 mm; 3, particle size 5.0 mm); 4, 5, 6, impact stressing (4, particle size 0.08 mm; 5, particle size 0.54 mm; 6, particle size 6–7 mm).

feed whose particles are small and not too hard. In these cases the daughter particles generally remain where the parent particle was and can be briquetted as they continue to be stressed. A tangential movement of the milling implement distributes the fragments to a greater extent over the surface and thus lessens the proportion of agglomerates in the product. Table 3.1 shows this effect for particles of limestone and cement clinker

Table 3.1 Percentage by mass of agglomerates produced in the crushing of lime-stone and cement clinker with and without friction in a roll mill (rounded off figures)

Reduction ratio		2		4		8	
Ratio of rotational speeds		1:1	2:1	1:1	2:1	1:1	2:1
Limestone	200 μm	1	0	11	0	42	4
	500 μm	0	0	4	0	21	2
	1000 μm	0	0	2	0	8	1
Cement clinker	200 μm	1	0	6	0	21	1
	500 μm	0	0	2	0	8	0
	1000 μm	0	0	1	0	4	0

crushed between two rollers. The ratio of the rotational speeds was either
1 : 1 or 2 : 1. Without friction the formation of agglomerates is larger for
the soft limestone than for the hard cement. It occurs to a greater extent
when the particles are small and the reduction ratios are larger, and is
drastically reduced by friction [71].

Based on the laws of fracture mechanics the following properties of the
utilization of energy S_E can be derived by dimensional analysis [80].

1. For physically identical materials S_E is constant if the product $E_v x$ is
   kept constant and if it is assumed that the shape of the particles is
   similar, including the distribution of cracks and the states of stress and
   strain before fracture begins, and that grain deformation is purely elastic.
   This principle of similarity thus leads to Rittinger's law, but only under
   the assumptions given above.
2. For particles of the same material which, however, have differing
   crack-length distributions

$$S_E \beta_{max} = f\left(\frac{E_v x}{\beta_{max}}, \frac{l_i}{x}\right) \tag{3.67}$$

where E_v is the energy of comminution per unit volume and l_i is a
characteristic crack length. The maximum specific fractured-area energy
β_{max} is a property of the material. Thus S_E is independent of the particle
size x only if $E_v x$ and l_i/x are constant. The graph of S_E versus x with
$E_v x$ as the parameter is useful in explaining how a fracture proceeds.

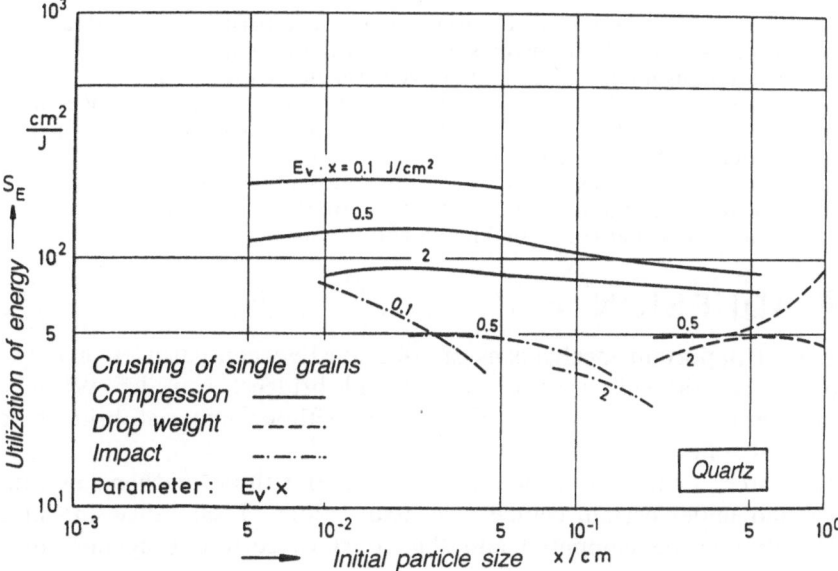

Figure 3.34 The utilization of energy, for constant energy input $E_v x$, as a function
of initial particle size in the crushing of quartz by a drop weight, compression and
impact.

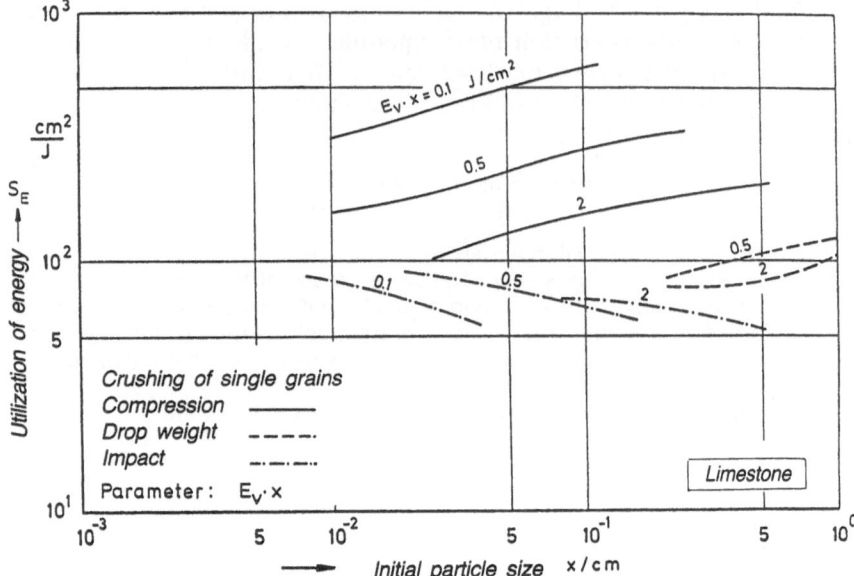

Figure 3.35 The utilization of energy, for constant energy input $E_v x$, as a function of initial particle size in the crushing of limestone by a drop weight, compression and impact.

Corresponding results for experiments on single particles are plotted in Figs 3.34 and 3.35. In the crushing of quartz by compression the assumptions made appear to be largely satisfied and, to a good approximation, the plots are horizontal. However, deviations above and below horizontal are observed when quartz is stressed by impact or under the action of a dropped weight and in all cases for limestone. These deviations can be explained by agglomeration processes and the dependence of the crack size distribution on the particle size x. This has been discussed at length by Pietsch and Rumpf [75].

3.6 ADHESION

The most important mechanisms of adhesion between particles or between particles and a wall arise through liquid bridges, van der Waals and electrostatic interactions, which also act without material bridges, and bridges of solid matter.

The forces of attraction operating in the first three mechanisms can be calculated under certain geometrical assumptions. Their range and magnitude will now be compared and the effect of surface roughness on the adhesion will also be discussed.

Section 3.6 was written with the cooperation of Messrs S. Moser and K. H. Sartor.

3.6.1 Liquid bridges

If liquid is present between the surface of two solid bodies, boundary forces arise as a result of the surface tension γ of the liquid which always has an attractive effect on the surfaces of the bodies. Moreover, as a general rule a difference in pressure, called the capillary pressure, is established. The capillary pressure can be calculated from Laplace's equation (2.48). In liquid bridges this pressure is constant if the effect of gravity is negligible:

$$p_c = \gamma \left(\frac{1}{R_1} + \frac{1}{R_2} \right) = \text{const} \tag{2.48b}$$

R_1 and R_2 are the principal radii of curvature of the liquid surface.

If the pressure within the liquid bridge is less than the external pressure, p_c is defined as positive. The capillary pressure then adds to the adhesion. If the capillary pressure is negative the resultant effect is repulsive. The adhesive force always consists of boundary forces and capillary forces.

> The conditions that have to be satisfied by the forces acting in a liquid bridge between two equally large spheres lead to a linear differential equation of the first order with elliptical integrals for which there is a closed solution only in certain cases. Schubert [24, 85] has used numerical methods to determine the whole range of solutions for a liquid bridge between two spheres. Results for any system of rotationally symmetric bodies such as two spheres with different diameters as well as a sphere and a plate and a cone and a plate are also available [86].

In the sphere–sphere model the liquid bridge is determined by the surface tension γ of the liquid, the diameter x of the spheres, the half-angle β subtended at the centre, the angle of contact δ and the gap a between the surfaces of the two spheres. The following dimensionless expressions are obtained for the adhesive force H, the capillary pressure p_c and the volume V_b of the liquid bridge: the relative adhesive force is given by

$$\frac{H}{\gamma x} = F_H \left(\beta, \delta, \frac{a}{x} \right) \tag{3.68}$$

the relative capillary pressure is given by

$$p_c \frac{x}{\gamma} = F_{p_c} \left(\beta, \delta, \frac{a}{x} \right) \tag{3.69}$$

and the relative liquid volume is given by

$$\frac{V_b}{2\pi x^3 / 6} = \phi \left(\beta, \delta, \frac{a}{x} \right) \tag{3.70}$$

Table 3.2 Relative adhesive force F_H for liquid bridges between two spheres of equal size

a/x	δ	β	F_H
0	$\geqslant 0$	> 0	$\approx \pi \cos \delta \{1 - \beta/(\pi - 2\delta)\}$
0	$\geqslant 0$	$\to 0$	$\approx \pi \cos \delta = F_{H.\max}$
0	0	> 0	$\approx \pi - \beta$
> 0	$\geqslant 0$	0	0
$\geqslant 0$	$\geqslant 0$	> 0	$= \pi \sin \beta \{\frac{1}{4} F_{p_c} \sin \beta + \sin (\beta + \delta)\}$
0	0	$57°$	$= 2.2 \qquad F_{p_c} = 0$

where $2\pi x^3/6$ is the volume of solid matter in the two spheres. The function F_H (eqn. (3.68)) is given in Table 3.2.

The values of the functions $F_H(\beta, \delta, a/x)$, $F_{p_c}(\beta, \delta, a/x)$ and $\phi(\beta, \delta, a/x)$ have been estimated numerically and recorded in the form of diagrams [24, 85]. If, for example, the relative adhesive force F_H is plotted against the reduced interval a/x for a constant liquid volume ϕ the stress–strain diagram for the liquid bridge is obtained [24].

3.6.2 van der Waals forces (dispersion forces)

The interactive energy for the van der Waals attraction is of the order of 0.1 eV and decreases with the sixth power of the distance between the molecules. The range of the van der Waals interaction is large compared with that of the chemical bond.

Two different theoretical paths have been used to calculate the attractive forces between two bodies. Hamaker's microscopic theory [48] is based on the assumption of the additivity of the forces between the molecules of both bodies. With this theory the attractive force H can be calculated for various geometries, e.g. the result for the system consisting of a sphere and an infinite semi-space is

$$H = \frac{AR}{6a^2} \tag{3.71}$$

where a is the gap between the surface of the sphere and the semi-space (plate). When these adhere a becomes equal to a_0 with a value of about 4 Å [59]. R is the radius of the sphere and A is the Hamaker constant which can only be roughly determined for most materials (order of magnitude about 10^{-12} erg).

In Lifshitz's macroscopic theory [65] the dipole interaction is considered as a property of the electromagnetic field. The system minimizes its energy by phase correlation. The van der Waals attraction is the result of this phase correlation. According to Krupp [59] the relation obtained for the sphere–plate system is

$$H = \frac{\hbar\bar{\omega}R}{8\pi a^2} \qquad (3.72)$$

where $\hbar$ is the Lifshitz–van der Waals constant and $\bar{\omega}$ depends on the dielectric constants of the two bodies. The product $\hbar\bar{\omega}$ can thus be calculated from known macroscopic data for the adhering pair of objects. It lies between 0.1 and 10 eV.

3.6.3 Electrostatic forces

When two solid surfaces come in contact electrostatic forces of attraction arise as a result of the contact potential thus generated. Surplus charges can also cause attractive or repulsive forces. Such charges arise, for example, through the frequent collisions of particles with one another or against boundary surfaces (contact charging) whenever one of the colliding bodies is an insulator (no discharge on separation).

The physical cause for the transfer of electrons when two bodies come into contact is the difference between their electron work functions. Electrons migrate from the body with the smaller work function to the body with the larger work function. The equilibrium charge is then reached when the potential difference caused by the different work functions is equal to the potential difference due to the separated charges. For conductors, semiconductors and insulators the contact charge is proportional to the difference between the electron work functions of the pair of bodies in contact. Moreover, in the case of semiconductors and insulators, the density of electrons on the surface has a strong influence on the contact charge [17]. This charge is less than about 10^{10} e/cm^2 for insulators, between 10^{11} and 10^{12} e/cm^2 for semiconductors and over 10^{12} e/cm^2 for conductors. With conductors the charge is limited only by the capacity of the system.

The force of attraction between two charged surfaces can be calculated from the energy W of the electrostatic field. For conductors separated by a small distance a ($a \ll x$, where x is the diameter of the sphere) it can be obtained by differentiating W with respect to a where $W = CU^2/2$ (U is the potential difference between the two surfaces and C is the capacity which depends on the geometry). The result for the sphere–plate system is [59]

$$H = \pi\varepsilon_0 R \frac{U^2}{a} \qquad (3.73)$$

For insulators any distance apart and widely separated conductors $(a \gg x)H$ can be calculated using Coulomb's formula.

3.6.4 Solid bridges

If solid bridges are formed between solid particles, e.g. by a sintering, melting or crystallization process, then the adhesive force can be calculated if the cross-sections of the bridges and the strengths of the bridge-forming materials are known. However, these factors are generally unknown and are difficult to measure. Moreover, the solid bridges are often porous and not completely filled with material; therefore it is better to rely upon actual measurements of the adhesion.

3.6.5 Comparison of the first three adhesive mechanisms and the influence of the surface roughness

A general picture of the order of magnitude of the first three mechanisms of adhesion is presented in Fig. 3.36 where the adhesive force H for the sphere–plate system is plotted against their separation a. van der Waals forces fall off very quickly as the distance increases and have almost vanished at $1 \mu m$. Liquid bridges come into force only when they are formed at a very small separation a and this separation is then increased. They disappear suddenly at a particular value of a/x. For $a > 1 \mu m$ essentially only the electrostatic forces of attraction remain; these are plotted for conductors and insulators (homogeneously charged sphere in relation to either an infinite semi-sphere or a small circular charged body).

For ideally smooth surfaces the adhesive forces due to liquid bridges and the van de Waals and electrostatic interactions increase linearly with the radius R of the sphere for the sphere–plate system. An exceptional case is that of insulators for which a constant charge density σ is assumed; here the force H increases quadratically with R (Fig. 3.37(b)).

Rough surfaces can be ideally represented by means of asperities with a radius of curvature r. Instead of eqn (3.72) for the van der Waals adhesion the relation that now applies is

$$ H = \frac{\hbar\bar{\omega}}{8\pi} \left(\frac{r}{a^2} + \frac{R}{(r + a)^2} \right) \tag{3.74}$$

The effect of roughness on the adhesive forces is shown in Fig. 3.37(a). The corresponding maximum values (smooth surface) can be read from Fig. 3.37(b). This effect is most pronounced in the case of van der Waals forces because of their short range. If fine particles of diameter 10^{-1}–$10^{-2} \mu m$, say, are found between large particles they cause the same reduction in adhesion as the asperities. This effect is used in practice to

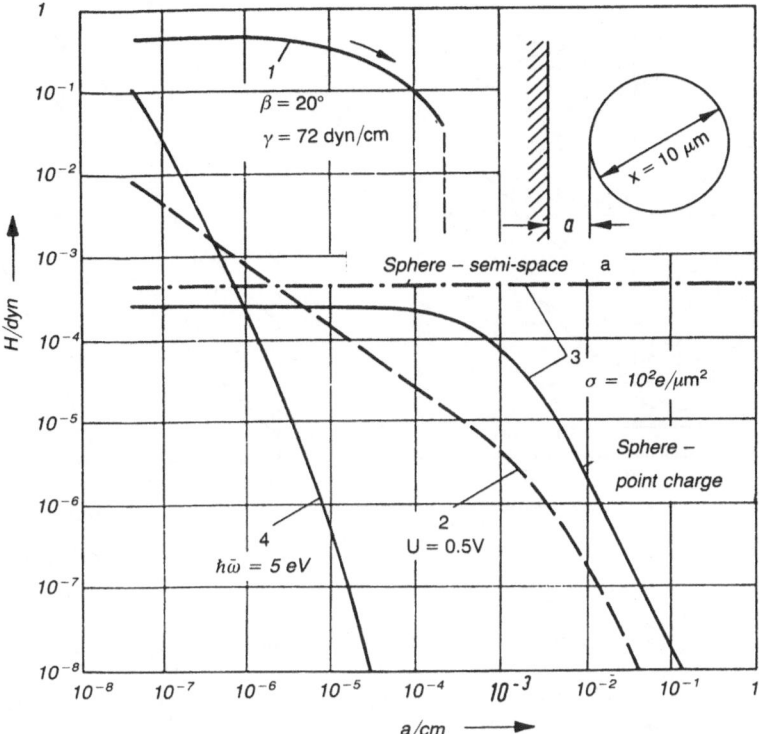

Figure 3.36 Adhesive forces due to various mechanisms of adhesion as a function of the distance *a* between the surfaces of a sphere and a plate: 1, liquid bridges; 2, electrostatic forces for conductors; 3, electrostatic forces for insulators; 4, van der Waals forces.

prevent agglomeration when, for example, fine-grained silicon dioxide (Aerosil®) is added to a mixture.

3.7 PARTICLE METROLOGY

Particle size measurement [1, 2, 8, 63] is the determination of the size distribution of particles (disperse phase), or a part or moment thereof, in a dispersion medium (continuous phase). Not only the disperse phase but also the continuous phase can assume any one of the three states of aggregation: solid, liquid or gas. Particle metrology, as perceived in mechanical process technology, is not concerned with molecularly disperse systems such as alloys, solutions or mixtures of gases.

Section 3.7 was written with the cooperation of Dr B. Koglin.

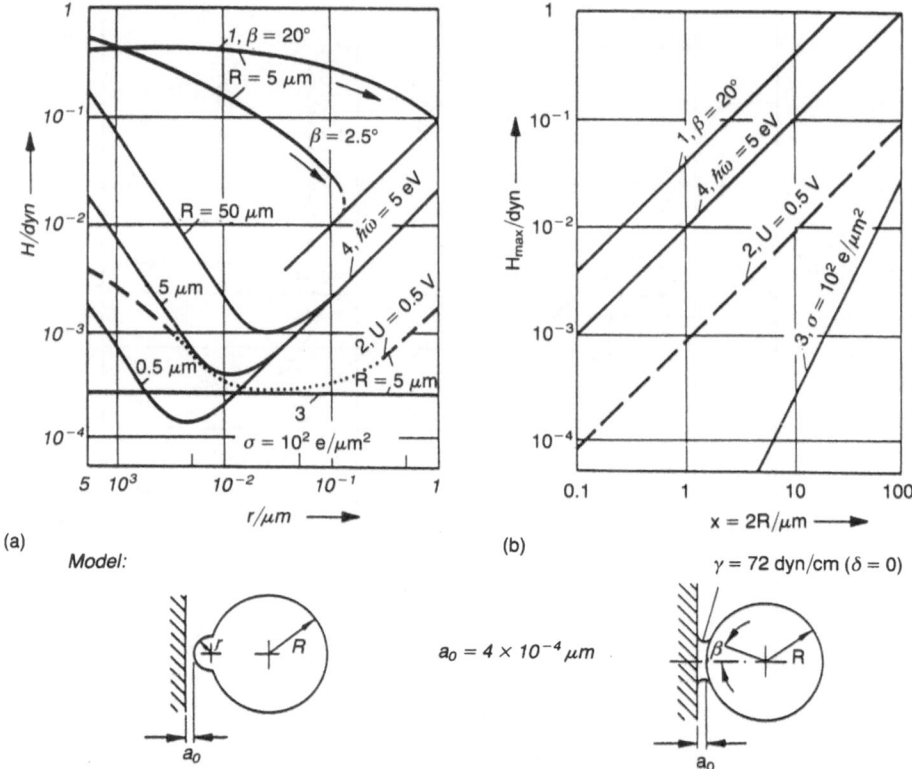

Figure 3.37 (a) Influence of the radius r of an asperity on the adhesive forces due to various mechanisms of adhesion for the sphere–plate model (contact gap $a = a_0$). (b) Adhesive forces of various mechanisms of adhesion for the sphere–plate model (on contact) as a function of the diameter of the sphere: 1, liquid bridges; 2, electrostatic forces for conductors; 3, electrostatic forces for insulators; 4, van der Waals forces.

Systems with a solid as the disperse phase and a liquid or a gas as the dispersion medium are of greatest importance. The investigation of these systems is called particle size analysis; it should be distinguished from grain analysis as understood in metallurgy and mineralogy, which is concerned with solid dispersions. Because of the growing application of nebulization involving liquids, measurements made on liquid phase dispersed in gaseous phases have also increased in importance (**droplet size measurement**). Liquids dispersed in a liquid phase (emulsions) and gas bubbles (bubble columns) involve a further branch of particle metrology. A special case is that of gases or liquids dispersed in a solid phase (pores). Here a distinction must be made between closed and open pores. The **porosimetry** of open pores was discussed in section 2.5.5.

The procedures for measuring particle size distributions yield the distribution function $Q_r(x)$ of some measure x of dispersity (fineness parameter) that depends on the geometrical dimensions of the particle. The large number of procedures available is due, on the one hand, to the numerous possible combinations of various fineness parameters and measures of populations and, on the other hand, to the various measuring techniques that involve the same principle of measurement. The most frequently used methods of particle size analysis are set out in Table 3.3 according to the dispersity parameter and the measure of population. In a practical but not strictly systematic classification of the methods of particle size analysis three types of method can be distinguished: counting procedures in which number is the measure of population, sedimentation methods in which the settling velocity is the dispersity parameter, and separation processes in which the material to be analysed is split into at least two fractions according to a given measure of dispersity.

Various moments of the particle size distribution are set out in Table 3.4. The moments are proportional to specific measures of the whole analysis sample, i.e. the length-, area-, mass- or volume-related number, length or area. Since the total mass of the material is not generally altered by the methods used in mechanical process technology, special significance is attached to the mass-related indices and also to the volume-related indices when the density is known and constant.

For narrow fractions of irregular particles the mass- or volume-related number, which is determined by the counting–weighing method, is important for ascertaining the equivalent diameter of the sphere with the same volume. The mass-related length is a common measure in the fibre industry. However, the mass- or volume-related surface area (specific surface) is by far the most important. Therefore, in the usual classification of particle size analysis, it is not generally the procedures for determining the moments of the size distributions but the procedures for measuring the surface area that are compared with the methods for measuring these distributions.

Only the external specific surface is proportional to the moment $M_{-1,3}$ of the particle size distribution; the shape factor is constant. The internal specific surface of porous systems is correspondingly related to the pore size distribution.

With some modifications the methods of particle size analysis for disperse solids can also be applied to some extent to drops and bubbles. The analysis of grain sizes and closed pores in a solid matrix is based on the microscopic analysis of sections. These methods of particle metrology fall into the category of counting methods. The relations between the distribution of particle cross-sections and the particle size distribution, i.e. the relations given on p. 48, are particularly useful for evaluating the results. The specific surface of the particles can also be calculated from the

Table 3.3 Classification of the methods for measuring particle size distributions

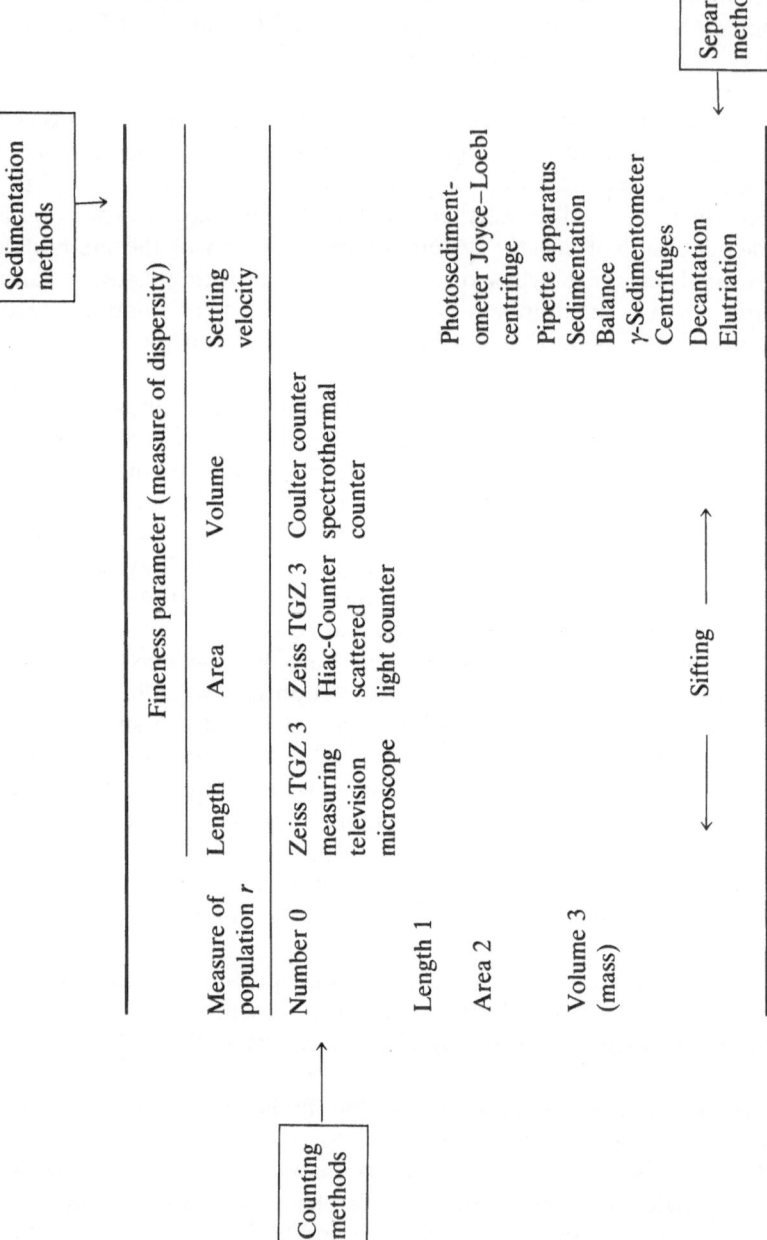

| | Fineness parameter (measure of dispersity) | | | |
Measure of population r	Length	Area	Volume	Settling velocity
Number 0	Zeiss TGZ 3 measuring television microscope	Zeiss TGZ 3 Hiac-Counter scattered light counter	Coulter counter spectrothermal counter	
Length 1				
Area 2				Photosedimentometer Joyce–Loebl centrifuge
				Pipette apparatus
				Sedimentation
Volume 3 (mass)				Balance
				γ-Sedimentometer
				Centrifuges
		Sifting ⟶		Decantation
				Elutriation

Counting methods →

Sedimentation methods →

→ Separation methods

Table 3.4 Systematic representation of the methods for determining the moments of a particle size distribution

Counting–weighing methods

	Per volume	Per area	Per length
Number	$\dfrac{M_{0,0}}{M_{3,0}} = M_{-3,3} = \dfrac{1}{M_{3,0}}$	$\dfrac{M_{0,0}}{M_{2,0}} = M_{-2,2} = \dfrac{1}{M_{2,0}}$	$\dfrac{M_{0,0}}{M_{1,0}} = M_{-1,1} = \dfrac{1}{M_{1,0}}$
Length	$\dfrac{M_{1,0}}{M_{3,0}} = M_{-2,3} = \dfrac{1}{M_{2,1}}$	$\dfrac{M_{1,0}}{M_{2,0}} = M_{-1,2} = \dfrac{1}{M_{1,1}}$	
Area	$\dfrac{M_{2,0}}{M_{3,0}} = M_{-1,3} = \dfrac{1}{M_{1,2}}$		

Methods for measuring the (external) specific surface: photometer, permeameter

All moments can be determined using graphical procedures (Zeiss TGZ 3, measuring television microscope). For the moments $M_{-n,3}$ the particle volume per image area must be known.

moments of the distribution of cross-sections.

In the following treatment of particle metrology the usual classification into methods involving counting, sedimentation, separation, surface area measurements and pore size measurements will be followed. Before doing this, however, it is necessary to discuss sampling and dispersing which are often larger sources of error and demand greater effort than the actual measuring procedure.

3.7.1 Sampling and sample splitting

The purpose of particle metrology is to ascertain, for a given parent particle population, its distribution function or a moment of either the whole or a part of the particle size distribution. As a general rule the parent population is much larger than that which can be subjected to measurement without excessive effort. Therefore it is necessary to take a sample for investigation and analysis.

The total error in the result for the parent population consists of the sampling error and the measuring error. In so far as these errors are of a statistical nature the sum of their squares is equal to the square of the total error, i.e. their variances can be added. The sampling error should not exceed the actual error of the measurement. As a reference value a statistical error of 2% can be cited, which experience shows must be assumed even when the analysis of particle size has been carefully carried out.

If there is no correlation between the various elements of the sample then the expected sampling error can be calculated from the number of individual particles in the sample. The statistical distribution of the properties of the sample is described by the hypergeometric distribution, which can be approximated by the binomial distribution when the size of the total population is large compared with the sample size. If the cumulative distribution by number Q_0 takes values close to 0.5, the binomial distribution can, in turn, be approximated by the Gaussian normal distribution (cf. section 2.2.5) when there are a large number of particles in the sample ($n \gtrsim 100$). Accordingly the problem is reduced to the problem of sample size dealt with in section 2.4.3 for the case where the varying composition of the sample is normally distributed. The number n of particles in the sample which are necessary to ensure that, for a given probability $H(t) - H(-t)$, the absolute error ΔQ_0 of the cumulative distribution by number Q_0 does not exceed a defined limit $\Delta Q_{0\,max}$ is calculated using

$$ n = \frac{t^2}{\Delta Q_{0\,max}^2} Q_0(1 - Q_0) \qquad \Delta Q_0 \leqslant \Delta Q_{0\,max} \qquad (3.75) $$

with the probability $H(t) - H(-t)$. For $H(t) - H(-t) = 95\%$, $t = 2$. The values of $\Delta Q_{0\,max}$ and n for $Q_0 = 0.5$ are given in Table 3.5.

It can be seen that the analysis sample must consist of some thousands of particles to ensure that the statistical sampling error remains below the desired limit of 2%. As a rule this condition has to be considered only when counting procedures are used or the material is coarse. With other methods of particle size analysis the number of fine particles in the analysis sample are almost always much higher. However, the necessary sample size increases substantially if correlations exist between different elements of the sample. Correlations can occur in the sample if samples are taken from a segregated parent population or if systematic errors are made when sampling a parent population that is a uniform random mixture. If the mass of the sample contains components that are fully correlated, i.e. segregated, then the number n in eqn (3.75) and Table 3.5 does not denote the number of individual particles but the number of mutually independent but internally fully correlated components.

Table 3.5 Required sample size n as a function of the maximum permissible statistical error $\Delta Q_{0\,max}$

$\Delta Q_{0\,max}$ (%)	0.5	1	2	5	10
n	40 000	10 000	2500	400	100

The measures that have to be taken to prevent a systematic sampling error depend on the nature and condition of the parent population and cannot be dealt with in a general way. In the case of a segregated parent population, which can rarely be ruled out, the principle holds that a laboratory sample, whose bulk exceeds that of the analysis sample, should consist of the greatest possible number of the smallest possible independent individual samples. For fine-grained materials the laboratory sample should amount to, say, 1 kg. The analysis sample is obtained from this sample by sample splitting which, down to amounts of about 10 g, can be done by coning and quartering or riffling. Spinning rifflers divide the laboratory sample into the largest number of individual elements (maximum about 10^4) and enable the best splitting of dry and free-flowing powders to be carried out. Splitting of samples of about 10 g or less can be carried out by pipetting a portion out of a suspension or by taking it from a well-mixed paste.

3.7.2 Dispersion

The particles to be investigated can be either individual particles or more or less large agglomerates of these. In taking the sample, preparing it for measurement and carrying out the actual measurement the degree of agglomeration can be altered. If the problem is to determine the sizes of the particles in a given system and for the degree of agglomeration that is characteristic of that system, the measurement must take place without disturbing the system in so far as the strength of the agglomerates, e.g. aggregated pigments, permits.

In many cases it is not the size of the particles occurring at a certain degree of agglomeration but the size of the individual particles that is of interest. In this instance care must be taken to disperse the individual particles in a fluid. Attraction between similar solid particles occurs in any fluid as a result of van der Waals forces. Furthermore, in the case of dry powder electrostatic forces of attraction frequently act over larger distances. In the case of dispersions of moist powders in gases capillary forces also contribute to the adhesion because of liquid bridges.

For the dispersion of solids in suspensions [16, 18] the choice of a suitable liquid is of fundamental importance. It follows from the macroscopic theory of van der Waals forces that the static dielectric constants of solid and liquid can be used for the approximate estimation of the forces of adhesion. If, on the one hand, the liquid has a large dielectric constant and, on the other hand, there is little difference between the dielectric constants of the solid and the liquid, then the adhesive forces are small and there is little agglomeration. These tendencies have also been confirmed experimentally. The electrostatic forces of repulsion between the suspended solid particles are determined by the zeta potential established at the liquid–solid interface. As a result of this, great importance is attached to the ionic surfactants which are added to the suspension in a small concentration as dispersants. The surface-active ion of these dispersants is adsorbed on the surface of the solid and can thereby build up a large zeta potential. Of course the electrostatic potential near the surface is diminished by the diffuse layer of oppositely charged ions, and hence the repulsion is reduced.

The dispersion of dry solids in gases can be facilitated by compensating charges, e.g. by ionizing the gas. However, the addition of dispersants that carry electric charges can also achieve the same result. Highly dispersed silica with particles of dimension about 10^{-2} μm (e.g. Aerosil®) has proved to be particularly useful for this purpose; mixed with the powder in small amounts it is adsorbed on the surface of the solid which is thereby covered uniformly with negative electric charges. At the same time, because of the small particle size, it considerably reduces the van der Waals attraction, often by several orders of magnitude (cf. p. 118).

3.7.3 Counting methods

In counting methods [30] the measure of dispersity, i.e. the particle size, is ascertained for each individual for each individual particle. The smallest number of particles to be measured is determined by the permissible statistical sampling error (cf. p.000). The requirement to analyse well over 1000 particles in as short a time as possible has led to the rapid development of automatic counting methods in recent years.

(a) Indirect (pictorial) counting methods

In the case of direct counting methods the particles are generally viewed through a microscope. The simplest devices for analysing the image are ocular micrometers, in which the particle is compared with a linear scale, and ocular graticules, in which it is compared with a circle. With the image-shearing eyepiece the image of the specimen is split into two images which can be moved relative to each other by a known amount. If the microscope image is projected onto a frosted glass screen, or a photograph is taken, the picture can then be compared with scales or patterns.

The analysis of the photographs can be performed much more rapidly using the Zeiss TGZ 3 *Teilchengrößenanalysator* (particle-size analyser). The image of an iris diaphragm illuminated from below is projected onto the surface of a clear Plexiglas plate on which the slightly transparent photograph is laid. Superimposed on the photograph is a sharply defined circle of light whose area, by adjustment of the iris diaphragm, can usually be made to agree with the projected area of the particle being measured. The circumscribing and inscribing circles and the largest chord can also be measured. When the iris diaphragm is correctly adjusted a counter allocated to the diaphragm diameter is actuated by a foot-operated switch. At the same time the picture of the particle is perforated with a sharp marking device in order to avoid repeated counting. Finally the next particle image is moved over the circle of light. About 2000 particles per hour can be measured in this way.

If an image is to be automatically analysed it must be transformed into a sequence of signals. The transformation is effected by line-by-line scanning of the image with a light-sensitive detector. A distinction is made between scanning with a point, a line or a surface. In recent years various instruments have been developed in which the image analysis is carried out using television equipment: the Quantimet (Metals Research), the Classimat (Leitz), the Micro-Videomat (Zeiss) and the PiMC Particle Measurement Computer System (Millipore). In these instruments the electrical signal from the television camera tubes is first fed to a pulse-height discriminator whose adjustment is determined by the brightness of the particle images and the background. Then the picture from the pulse-height discriminator, which in the first instance contains the information about the surface area of the particle image as a fraction of the whole field of view, is fed to a pulse-duration discriminator which in turn feeds to a store only pulses longer than a minimum duration corresponding to a minimum chord length in the direction of the scanning lines. By means of a grating each new signal is compared with the signal from the preceding line. If the signals from two neighbouring lines overlap then they are recognized as belonging to one particle. Counting is first initiated at the lower edge of the picture of the particles when no signal corresponding to a stored signal follows in the next line. The number of particles as a function of the threshold of the pulse duration gives the cumulative distribution by number as a function of the longest chord in the direction of the scanning lines. In addition to the fraction of the area covered by the particles and the number of particles, the number of chords can be ascertained from the number of bright–dark transitions. Further results are calculated from these three values.

An extension of quantitative image analysis on the basis of mathematical morphology is exemplified by the Leitz texture-analysis system in which information about the image, independent of orientation, is obtained through yes–no transformations based on the choice of a suggested structural element.

(b) Direct counting methods

In all direct counting methods, except the direct measurement of coarse particles by hand, the particles dispersed in a gas or liquid are transported through a measuring device in which they trigger a signal that depends on their size. If the signal is not electric it is transformed into an electric signal and subsequently discriminated and counted.

In the Coulter counter (Coulter Electronics) and various other instruments based on the same principle a suspension of particles in an electrolyte flows through a counting orifice through which an electric current also flows. If a particle passes through the orifice the electrical resistance is increased and consequently there is a drop in potential. Under certain assumptions the pulse height is almost proportional to the volume of the particle.

In a continuous-flow photometer a suspension of the particles in an optically clear liquid passes through an illuminated enclosure. The reduction of the light level is proportional to the extinction cross-section of the particle which is different from the geometrical projected area because the extinction coefficients depend on the optical properties. The HIAC Solid Automatic Particle Counter is a commercially available counter based on this principle.

In a light-scattering counter the particles dispersed in a liquid or gas are illuminated in a sensing volume and the intensity of the light scattered through a defined angle is measured. In the scope of geometrical optics and for white light, with particle sizes above about $2-10 \ \mu m$, the intensity of the light is proportional to the square of the particle's diameter; in the domain of Rayleigh scattering and for white light, with particle sizes below about $0.1-0.5 \ \mu m$, the intensity of the light is proportional to the one-sixth power of the particle's diameter. In the intermediate domain of Mie scattering the intensity of the light oscillates as a function of the particle size. Light-scattering measuring devices must be calibrated.

Other counting devices use interference with the field in the resonance space of a laser (Schleusener's laser counter), interference with gas flow (Langer's acoustic counter) or the thermal emission of light from heated particles (spectrothermal counter).

3.7.4 Sedimentation methods

In sedimentation methods the terminal settling velocity of a particle in a quiescent fluid, usually a liquid, under the action of gravity or centrifugal force serves as a measure of its size, i.e. fineness. Since the shape of the particle is generally unknown, it is not possible to derive a geometric dimension from its settling velocity. Instead, the equivalent settling rate diameter is calculated using the appropriate resistance law, usually Stokes law, which applies for Reynolds numbers below 0.25. A distinction is made between the **suspension method**, in which the particles are initially uniformly dispersed, and the **superimposed-layer technique**, in which the sample is initially concentrated in a thin layer over the liquid in which it is to settle.

The settling velocity depends on the concentration. With solid concentrations by volume in excess of about 10^{-3} the mutual hydrodynamic influence of the particles can lead to a microscopic demixing of the suspension by giving rise to particle complexes that have a higher settling rate (cf. section 3.1.3(b)). With solids concentrations by volume above 10^{-1} the reduction in the settling rate is predominantly caused by the upward flow of the sedimentation liquid in accordance with the principle of continuity. In order to keep errors due to the interaction between the sedimenting particles as small as possible their concentration should be at the lower limit of measurement. In the instruments commonly used to investigate particles of sizes in the range below about 50 μm the influence of the walls of the vessel on the sedimentation is negligible. In principle the range of application of sedimentation analysis is not restricted to small particles. The lower limit is determined by the Brownian motion and the duration of the analysis. Particle sizes down to just under 1 μm can be measured under gravity and sizes down to about 10^{-2} μm can be measured in a centrifugal field.

To establish the size fraction as a function of the settling velocity either the solids concentration at a given height in the suspension (incremental methods) or the quantity of solids present below or above a given level (cumulative methods) is determined. The most common methods are the pipette method, the photosedimentation technique and the gravimetric method using a balance with a pan immersed in the suspension.

In the pipette procedure samples are withdrawn by pipette from a given level at predetermined times, the liquid is evaporated off and the solid residue is weighed. In order to achieve a sufficiently accurate measurement of the amount obtained the initial solids concentration must be about 0.2 vol.%. With the photosedimentometer the estimate of the solids concentration is based on the attenuation of a beam of light at a given depth. Because of diffraction the attenuation of the light is not proportional to the geometric projected area of the particles. The deviation from this is allowed for by the extinction coefficient which is a function of the particle size; in addition, the coefficient depends on the optical arrangement as well as the spectral distribution of the light and the sensitivity of the detector. In the photosedimentometer the solids concentration is usually about or below 10^{-4}. With the sedimentation balance the amount of sediment that has settled on the pan is weighed as a function of the time. The desired cumulative distribution curve can be obtained from a plot of these results by a single graphical or numerical differentiation. The solids concentration is governed by the sensitivity of the balance; the usual concentrations are about 10^{-3} by volume.

Superimposed-layer techniques are generally used for measuring particles under gravity with sizes above about 60 μm in the intermediate range between Stokes law and Newton's law [37]. The errors caused by density convection currents in this case are less than in the case of smaller particles because the settling velocity is faster than the convection current. With a cumulative procedure such as that of the sedimentation balance the cumulative distribution curve can be determined directly by means of the superimposed-layer method.

The superimposed-layer method is particularly useful for the sedimentation analysis of fine particles in centrifuges. Whereas with suspensions the flow patterns that occur when the centrifuge is started represent a source of error, with the superimposed-layer method we can wait until the solids-free liquid is rotating like a rigid body. A commercial centrifuge with which the superimposed-layer method can be used is the Joyce–Loebl disc centrifuge. A layer of the suspension is added on the inside to the rotating peripheral annulus of solids-free liquid. After a predetermined sedimentation time an inner portion of the liquid annulus which contains only the fine materials below a certain particle size is separated off. The outer portion of the liquid containing the coarse material can be pumped off into a storage vessel after the centrifuge has stopped.

3.7.5 Separation processes

Under the heading 'separation processes' we include those procedures of particle metrology in which the feed is split into at least two fractions that ideally contain only particles below or above a certain size (cf. sections 2.3 and 4.1). Real separations are not sharp but are characterized by grade efficiency curves. The problem with analytical separation processes is to fix and determine the cut point. With separation processes the measure used for the population of particles can be freely chosen since the fractions are obtained by preparation. As a general rule the mass of the particles is used.

The most important separation processes in addition to sedimentation methods, which can also be considered in this group, are sifting and air classification.

(a) Sifting

In sifting, the solid material to be analysed is separated either into two fractions using a single sieve or into several fractions using a nest of sieves whose apertures decrease from the top to the bottom. Apart from sifting by hand, a distinction is made between dry sifting in a mechanical sieve shaker, air-jet sifting where the material is lifted off the sieve by a stream of air and repeatedly transported to another place on the sieve, and wet sifting where the solids are poured onto the sieve in a suspension. As well as the usual sieve cloths there are photomechanically manufactured sieve plates. With air-jet sifting these can be used down to apertures of about 15 μm and with wet sifting down to 5 μm.

Sieves have a fairly wide distribution of aperture sizes. This has the effect that the cut point shifts towards a coarser level as the sifting time continues. The cut point for the sifting procedure being considered is determined after the completion of the sifting by obtaining as undersize a narrow fraction of the residue on the sieve from which, by using the counting and weighing procedure together with a knowledge of the density of the material, the diameter of the equivalent volume can be determined.

(b) Air classification

In a stream of air the feed material is separated by the action of gravitational and inertial forces, on the one hand, and of drag forces, on the other. The measure of size is the settling velocity corresponding to the Reynolds number at which the separation is effected. A distinction is made between counter-current classification (equilibrium classification) and cross-current classification depending on whether the coarse material in the sifting zone moves against or across the direction of the stream of air.

> Among the gravity air classifiers the Gonell-Sifter (Chemisches Laboratorium für Tonindustrie) and the Analysette 8 (Fritsch) work according to the counter-current principle. The Zig-Zag Sifter Multiplex (Alpine) and the Centrifugal Air Classifier Zig-Zag Sifter 100 MZR (Alpine) use a combined counter-current cross-current arrangement in the individual steps and, over all the steps, the equilibrium technique. Among other commercial centrifugal sifters the Bahco Classifier (Etablissement Neu) and the Holderbank Classifier (Gebrüder Bühler) make use of the counter-current principle.

3.7.6 Techniques for measuring surface area

Only in the case of geometric bodies is the surface area an unequivocally defined magnitude. In the case of real bodies quite different values of the surface area can be specified depending on whether and how exactly their roughness is taken into account. The finest resolution that is meaningful with respect to the determination of surface area is represented by the lattice spacing of solids and consequently by the size of atoms or molecules.

The external surface, say the surface of the convex envelope surrounding the particle, and the inner surface of microstructures and pores down to molecular dimensions are of quite different significance not only for technical applications but also for particle metrology. For example, the external surface area is increased by comminution; however, in general the internal surface area is barely affected by any process of comminution. The overall surface area, which is the sum of the internal and external areas, can be greatly increased by crushing and grinding in the case of non-porous materials but can only be increased slightly in the case of very porous substances.

(a) Photometric surface measurement

In the photometric procedure the first stage is the determination of the volume-related extinction cross-section A_v of the solid matter in a uniformly mixed suspension with a solids concentration by volume of the order of $C_v \approx 10^{-4}$. A_v is calculated from the transmission T of a ray of light that has passed through a layer of thickness L in the suspension, where T is given by the Lambert–Beer law

$$T = \exp\left(-A_v C_v L\right) \tag{3.76}$$

The various photometer arrangements differ from each other essentially in the source of light and the angular aperture between the cell and the diaphragm of the detector. In order to obtain the smallest possible variations in the extinction coefficient with particle size the most suitable source is white light. A wide angular aperture has the advantage that for opaque particles larger than 10 μm the extinction coefficient equals unity. In this case the extinction cross-section is equal to the projected area which for convex particles, according to Cauchy's theorem, is equal to a quarter of the surface area. The disadvantage of a wide angular aperture is the dependence of the extinction cross-section on the translucence of the particles. A narrow angular aperture (e.g. 1°) has the advantage that the extinction coefficient is almost independent of the translucence of the particles since no significant portion of the scattered light is picked up by the detector. However, with a narrow angular aperture the extinction coefficient is a complicated function of the particle size. For this reason the projected area, and consequently the surface area, can be ascertained only approximately from the extinction cross-section.

(b) Permeability methods

In permeability methods a fluid, usually a gas, is made to flow through a packing of the powder to be investigated (cf. section 3.2). The specific surface can be deduced from the resistance to flow. Because neither the influence of the size distribution and shape of the particles nor the influence of porosity on the resistance to flow through the bed is fully understood as yet, considerable errors (perhaps as high as a factor of 2) must be expected when determining the surface area of different kinds of samples. However, these methods are well suited to comparative tests on similar samples.

The analysis of the permeability measuring procedure depends on the Knudsen number, i.e. the ratio of the mean free path of the gas molecules to the dimensions of the pores in the packing. In the range of continuum flow (Knudsen number very much less than unity) the internal surface contributes almost nothing to increasing the resistance to flow. With air at atmospheric pressure this kind of flow prevails for $S_v \leqslant 6000$ cm^{-1}, corresponding to an average particle size of more than about 10 μm. In this case the Carman–Kozeny equation is generally used to analyse the result. The best known commercial instruments based on this principle are the Blaine Fineness Tester and the Fisher Sub-sieve Sizer.

In contrast, the resistance to flow through a packing of porous particles in the range of molecular flow (Knudsen numbers much greater than unity) increases since the molecules can penetrate the inner pores of the particles. In this case the internal surface is at least partly determined by the permeability procedure. A commercial Knudsen permeability meter, working with helium as the measuring gas at pressures below 20 torr, is manufactured by the firm Micromeritics.

(c) Sorption methods

The basis of sorption methods is the fact that the amount of a gas, or a dissolved or suspended material, adsorbed on an interface under specified conditions is proportional to the interfacial area presented. If the amount of sorbate necessary for monolayer coverage can be determined and if the area occupied by an adsorbed molecule or particle is known, the extent of the surface area can be calculated.

The adsorbed quantity can be determined either directly on the sample by means of a microbalance or from the change in the amount present in the surroundings of the sample. With gas sorption this is done by measuring the change in pressure or the concentration change in a carrier gas. In the case of sorption of dissolved or suspended substances this is done by measuring the change in their concentration in the liquid.

The sorption process required for the measurement can be induced by changes in temperature, pressure or concentration. Although the adsorption procedure (lowering the temperature or rising the pressure or concentration) is more often used to measure the surface area, desorption (raising the temperature or lowering the pressure or concentration) is also of importance.

The most commonly used sorption method is gas sorption evaluated using the adsorption isotherm derived by Brunauer, Emmett and Teller (BET) using simplifying assumptions; this isotherm gives the amount n of gas adsorbed relative to the monolayer amount n_m as a function of (a) the gas pressure p divided by the saturation vapour pressure p_s and (b) a constant C that depends on the adsorption energy. The expression for this function is

$$\frac{n}{n_m} = Cx\{(1 - x)(1 - x + Cx)\}^{-1} \tag{3.77}$$

where $x = p/p_s$. The BET equation has generally been found to be very useful for physical adsorption in the pressure range $0.05 < p/p_s < 0.35$.

If only the specific surface but not, say, the pore size distribution has to be determined from the sorption measurements, it is usually satisfactory to determine just a single point near the upper limit of validity of the BET isotherm. Since the constant C is generally much greater than unity (in most cases $C > 60$), $1 - p/p_s$ can be neglected compared with Cp/p_s, and the BET adsorption isotherm is reduced to the relation

$$n_m = n\left(1 - \frac{p}{p_s}\right) \tag{3.77a}$$

This or a slightly modified equation forms the basis for the following single-point BET surface-measuring instruments: Areameter (Ströhlein), Areatron (Leybold–Heraeus) and Micromeritics Model 2200.

3.7.7 Pore size measurement

Pores in solids vary considerably not only with respect to shape but also with respect to size. The measurement of pore size distribution is generally based on grossly simplifying assumptions. It is usually assumed that the pores are cylindrical. It is customary to subdivide the pores into three classes according to size: pores with radii below about 1.5 nm are called micropores, those with radii above about 0.1 μm are called macropores and those in the intermediate range are called transitional pores. The common methods for determining pore size are based on sorption measurements on porous solids as well as on the measurement of capillary pressure.

(a) Pore size distribution using sorption methods

In order to determine pore size by means of sorption methods the t curve for the material system under consideration must be known. This curve represents the statistical thickness t of the adsorbed film as a function of the relative pressure. Since this thickness t depends on the energy of adsorption, t curves must be ascertained for all the gas–solid combinations in question. They are determined for the particular gas–solid system by measuring the sorption isotherms on non-porous samples of known surface area.

In order to determine micropores the volume of the adsorbed layer is plotted against the statistical layer thickness t. Although for non-porous substances this gives a straight line through the origin whose gradient is a direct indication of the surface area, in the case of microporous materials the gradient decreases to the extent that the pore walls are eliminated as adsorbing surfaces because the pores have filled up. The size distribution of the micropores is ascertained from the decrease in the slope of the plot of adsorbed volume against the statistical layer thickness t. The limits of this method are fixed at the lower end by the diameter of the molecule of the adsorbate (in the case of water, a pore radius of about 0.2 nm) and at upper end by the t values corresponding to pressures close to the saturation vapour pressure (in the case of benzene, about 1.6 nm).

The size distribution of the transitional pores can be determined by measuring the capillary condensation as a function of the ratio p/p_s of the gas pressure to the saturation vapour pressure. The Kelvin equation

$$\ln\left(\frac{p}{p_s}\right) = -\frac{2\sigma V \cos\theta}{rRT} \tag{3.78}$$

describes the relation between p/p_s and the surface tension σ, the molar volume V of the condensate, the contact angle θ between the condensate and the solid, and the radius r of the cylindrical capillaries in which condensation is established at equilibrium. Here R is the universal gas constant and T is the absolute temperature. Because of the multilayer adsorption on the walls of the pores the statistical thickness t of the adsorbed film must be added to the radius of curvature calculated from the Kelvin equation to obtain the capillary radius r. A problem exists in that the sorption isotherms in the range of capillary condensation generally exhibit hysteresis. As a general rule the desorption branch of the isotherms is used when calculating the pore size distribution by means of the Kelvin equation.

(b) The determination of pore size by capillary pressure

As was explained in section 2.5.5 porosimetry does not measure the pore size distribution but a pore distribution depending on the constrictions surrounding the pores. However, the results are usually analysed by substituting the pore system with a system of cylindrical pores of circular cross-section and determining the pore radius r from the intrusion pressure p by the capillary pressure formula

$$p = \frac{2\gamma}{r}\cos\theta \tag{3.79}$$

The best known application is mercury porosimetry. Within a pressure range of 1–2000 atm a measuring range for apparent pore radius of 7.5 μm to 3.75 nm is obtained based on the surface tension $\gamma \approx 480$ dyn/cm for mercury and the contact angle $\theta \approx 141°$, which is approximately the same for contact with most solids. With low-pressure porosimeters using pressures down to below 30 torr pores with radii up to about 200 μm can be measured.

Pore radii distributions in the range between 1 and 200 μm can also be determined from the depression of the vapour pressure of wetting liquids.

4

Processes

4.1 SEPARATION PROCESSES

The separation processes used in mechanical process technology can be subdivided according to their purpose. The following processes can be distinguished.

1. Sorting: separating a particle assembly according to the kind of material:
2. Classifying: separating a particle assembly according to particle size:
3. Filtration and sedimentation: separating a particle assembly from the carrier phase. Such a separation is never complete. The particles separated by these processes remain in the carrier phase but in a very much higher concentration. The filtered or clarified carrier phase often also contains particles that have not been removed. The carrier phase can be liquid or gaseous, and the dispersed phase can be solid, gaseous or liquid.

In all cases in which a solid disperse phase is separated according to any measure of particle size the separation can be represented by a grade efficiency curve (see section 2.3). Separation by elutriation or settling in a stream of fluid can be used not only for sorting particles but also for classifying and removing them. The hydrocyclone is a typical example of the use of the same device to sort, classify and remove particles. It is therefore advisable to group the separation processes according to their physical function. The design of the equipment also corresponds to this.

4.1.1 Separations in streams of fluids

In separations in streams of fluids the particles execute a movement relative to the fluid, generally as a result of inertial and buoyancy effects. The inertial effect may be due to the gravitational or centrifugal force, or to any other inertial force. Separations from fluids brought about by electrical or magnetic forces will be treated separately (see section 4.1.3). The simplest example of these processes is separation by sedimentation in

Section 4.1 was written with the cooperation of Dr H. Reichert.

a static fluid. This is used not only for particle size analysis, using gravitational or centrifugal forces (see section 3.7.4), but also for preparative separation from liquids by means of batchwise-operating solid-bowl centrifuges.

(a) Cross-current separation processes

The removal of particles etc. by sedimentation under continuous or semicontinuous operation involves cross-current separation. Here a suspension of uniformly distributed solid particles, droplets or bubbles flows into a sedimentation chamber in which the suspended phase separates out crosswise to the fluid's direction of flow as a result of field forces (or buoyancy).

Important technical applications of cross-current separation under gravity are the use of clarifiers in the purification of waste water and the separation of low-density liquids, e.g. the removal of petrol from water. As a general rule clarifiers are installed in series with aeration equipment for biological purification and with flocculation equipment for agglomeration and coagulation. Figure 4.1 shows the basic construction of a clarifier without its auxiliary equipment. Cross-current separation is obtained under centrifugal force in solid-bowl centrifuges, e.g. in the gas centrifuge for separating isotopes. The tubular centrifuge, which works by the overflow principle, is used to remove particles from liquids under high centrifugal acceleration (separation of submicroscopic particles). Normally a decanting centrifuge is used, i.e. a solid-bowl centrifuge which, with the aid of a screw discharge, can be operated continuously. In the case of cross-current classification, as opposed to cross-current separation, the material is not suspended in the inflowing carrier medium but is added separately (Fig. 4.2). Cross-current classification under gravity is used in the winnowing of grain. The cross-current principle does not require a field force. It is sufficient that the incoming stream of material possesses a certain initial velocity transverse to that of the stream of carrier medium (Figs 4.2(a)–4.2(c)). By this means the stream of particles fans out and can be split into coarse and fine material using a blade. Rumpf and Leschonski's method of cross-current jet sifting has been developed for both plane-symmetric and rotation-symmetric systems. When the throughput is continuous it is possible in theory to obtain ideally sharp separations using this method.

Figure 4.3 shows some experimentally obtained grade efficiency curves as a function of the throughput ratio $\mu = \dot{M}_f / \dot{M}_a$ ($\dot{M}_f$ is the mass flow rate of the feed and $\dot{M}_a$ is the mass flow rate of the air). The experiments were carried out using a planar arrangement. With a low throughput ratio the grade efficiency curve is very steep in the range $G = 0.1$–0.9. As the throughput ratio increases the stream of the carrier is deflected by the angular momentum of the feed material with the result that, in contrast with spiral sifting, the separation is shifted towards finer sizes.

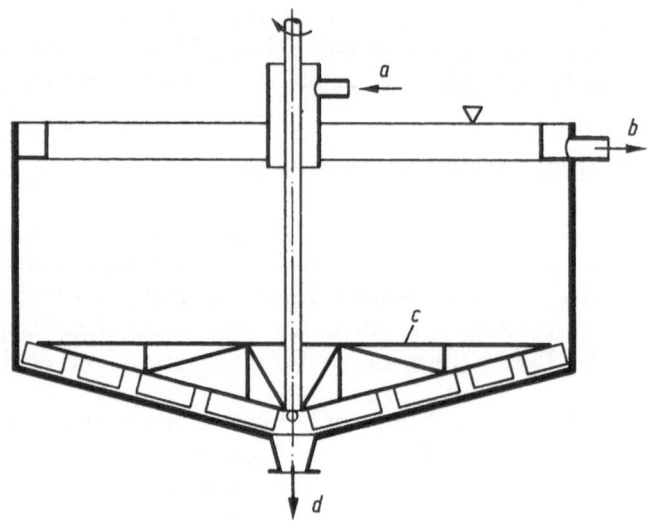

Figure 4.1 Schematic construction of a clarifier: a, suspension; b, clarified over-flow; c, rake; d, thickened underflow.

(b) Counter-current separation processes (equilibrium separation)

In the case of counter-current or equilibrium separation the stream of fluid flows in the opposite direction to the coarse material settling out. As a result of the equilibrium between the drag force and the field force (gravity or centrifugal force) particles of a certain size x_c, which is known as the cut size, will have an absolute velocity of zero. Theoretically such particles remain suspended. Coarser particles have an absolute velocity opposite to the direction of the stream and are removed as oversize. Their settling velocity is greater than the velocity of approach of the carrier fluid. Particles of size $x < x_c$ are carried away with the fluid and go into the fines. With batchwise feeding an ideally sharp separation can thus be theoretically achieved with a vertical grade efficiency curve at x_c. In practice batchwise feeding is used for particle size analyses in air classifiers operating under gravity (e.g. Gonellsifter and Analysette *et al.* (see p. 131)). In a shallow sifting zone an extremely sharp separation with a cut sharpness $\kappa \simeq 0.9$ can be achieved using the method of Rumpf and Leschonski (Fig. 4.4). In a centrifugal force field the principle is applied with a correspondingly sharp separation in the batchwise-operating spiral air classifier (Rumpf–Leschonski).

With continuous feeding the ideally sharp separation is fundamentally no longer possible. The concentration of particles of transitional, i.e. equilibrium, size would go on increasing without limit. Similarly the concentration

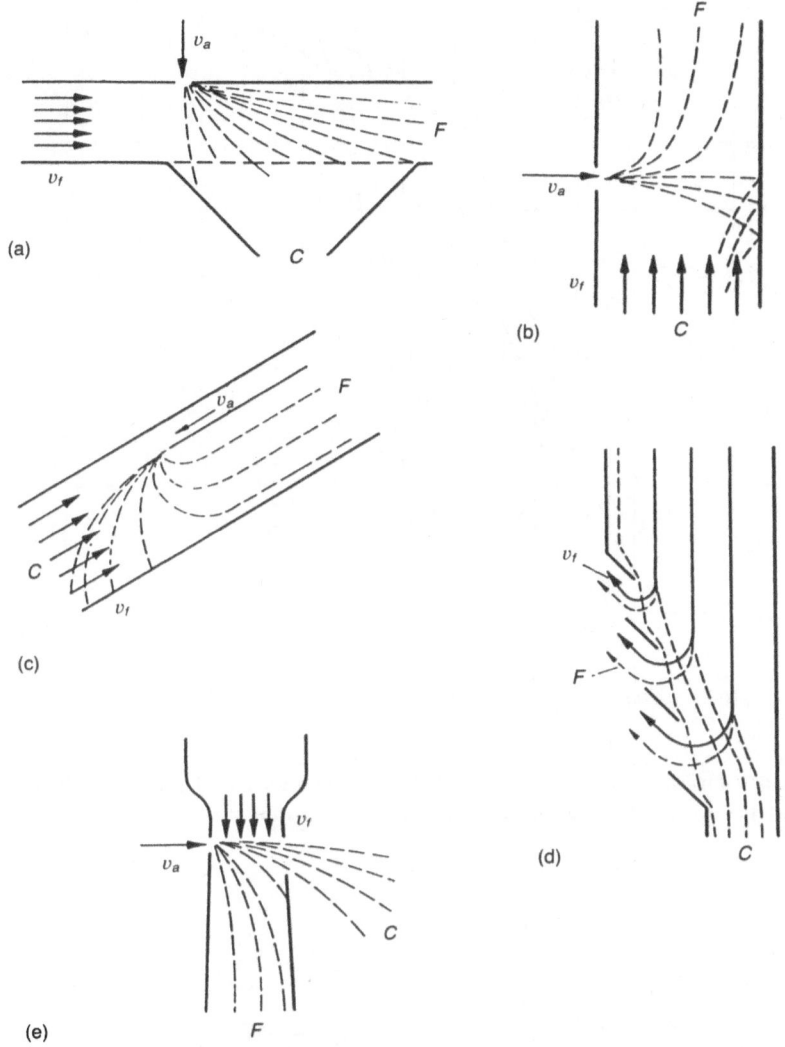

Figure 4.2 Cross-current classifying: (a) horizontal cross-flow classifier; (b) vertical cross-flow classifier; (c) counter-current elbow classifier; (d) louvre elbow classifier; (e) cross-current jet classifier; F, fines; C, coarse material; v_a, air velocity; v_f, feed velocity.

of adjacent particle size gradings would increase because of the slow rate at which they are discharged. In fact, concentration profiles appear for each grading. Because of the velocity of the fluid current a diffusion process, which, depending on the concentration gradient, distributes the particles on both the coarse side and the fine side, is superimposed on the

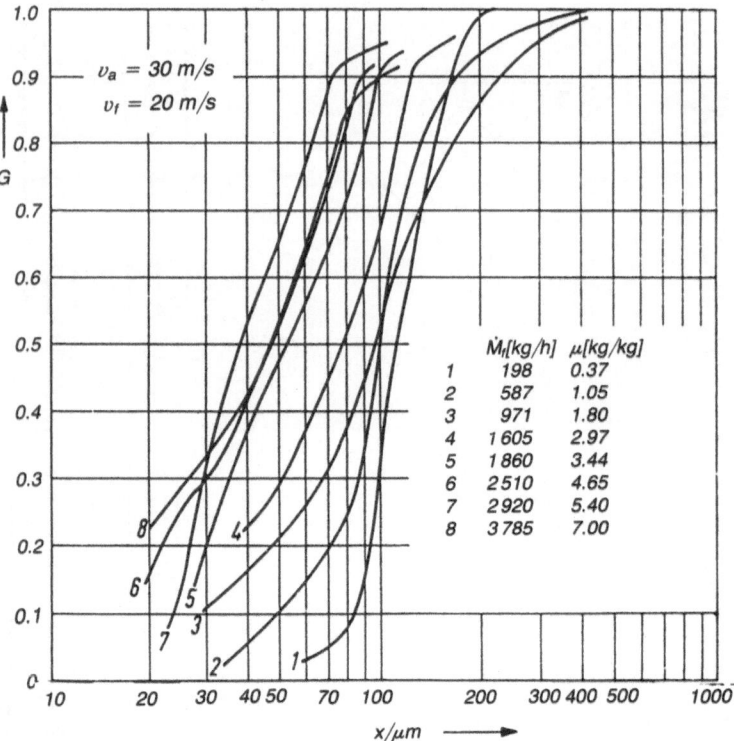

Figure 4.3 Grade efficiency curves $G(x)$ of the cross-current jet sifter showing their dependence on the throughput ratio $\mu = \dot{M}_f/\dot{M}_a$ employed.

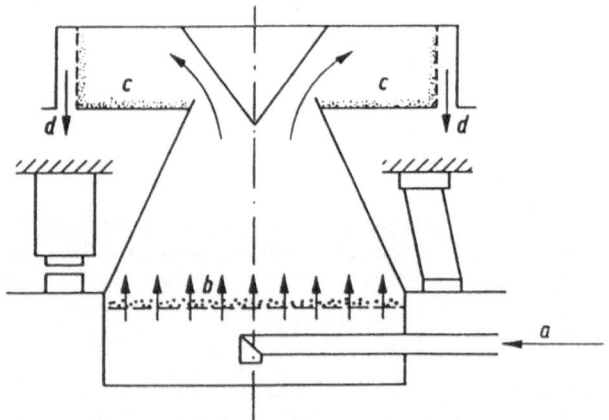

Figure 4.4 Sketch of the gravity air classifier with an extremely shallow cylindrical sifting chamber and accelerated removal of the fines: a, d, air; b, coarse material; c, fine material.

convective transport. In principle a calculable grade efficiency curve is thereby obtained which becomes less sharp as the feed or the throughput ratio increases.

In gases, because of the low discharge rate, a continuous counter-current separation under gravity would be applicable only in the case of coarse material. Other separation methods are used in this case. With liquids, counter-current separation under gravity is employed as in the classification of sand (Eder's elutriation method) for example.

In a centrifugal force field counter-flow equilibrium separation is achieved in the spiral air classifier [103]. The spiral classifier is a flat cylindrical chamber in which the carrier medium is fed in tangentially at the rim of the chamber and flows out axially at the centre. In the Walther classifier a boundary-layer stream which flows inward develops at the stationary walls. The considerable drop in throughput resulting from this arrangement is eliminated in the microplex sifter (Alpine) in which the walls rotate, thus supplying additional energy to the stream of fluid. The microplex sifter (Fig 4.5) produces sharp separations at cut points between 2 and 50 μm and throughputs between 20 and 5000 kg/h.

The counter-current principle has also been applied for a long time in the rejector-wheel classifier in which the particle-laden stream flows radially inwards through a rejector wheel fitted with vanes (Fig. 4.6). This classifier is frequently used in combination with mills.

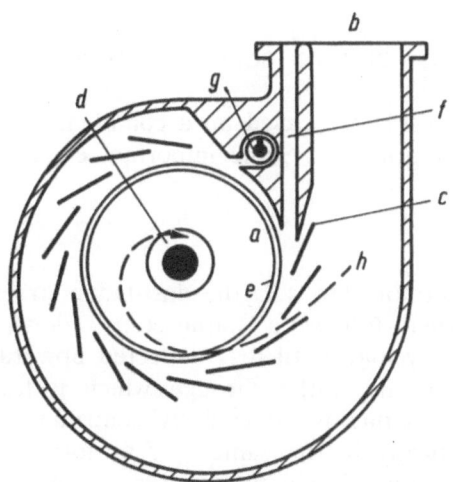

Figure 4.5 Microplex spiral air classifier (Alpine, Augsburg): a, sifting zone; b, air inlet; c, adjustable guide vanes; d, central exit of air and fines from the sifting zone; e, rotating walls of the sifting zone; f, feed inlet; g, screw for removing the coarse material; h, one of the spiral streamlines.

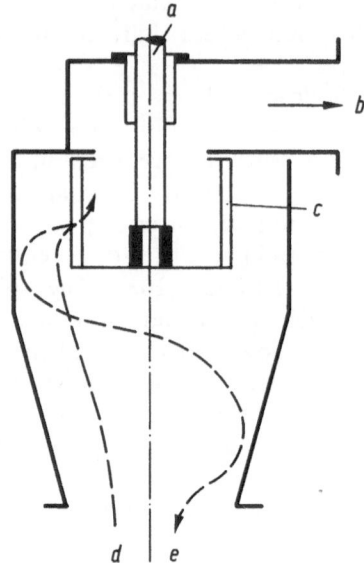

Figure 4.6 Rejector-wheel classifier for mounting on a mill; a, variable-speed drive; b, exit of fines and air; c, rejector wheel with vanes; d, entry for feed and air; e, recycle stream of oversize.

(c) Combined cross-current—counter-current separation processes

Most technical processes are based on a combination of the cross-current and counter-current principle. Some important examples are discussed in the following.

(i) The cyclone

In the cyclone separator (Fig. 4.7) the dust-laden stream of fluid flows in tangentially and then follows a spiral course downwards through the conical part of the cyclone until it is deflected upwards in the cylindrical zone beneath the round outlet through which it leaves the separator. Separation in the cyclone is effected by centrifugal forces and can be analysed mathematically in the same way as for plane flow in a spiral classifier. The separation is chiefly counter-current. The coarse particles are thrown out onto the wall of the cyclone and migrate in spiral strands downwards into the collecting hopper. The fine particles are carried away with the air through the outlet (vortex finder). The axial component of the flow superimposes the effect of cross-current flow on the counter-current separation. This results in particles which are rather larger than the cut

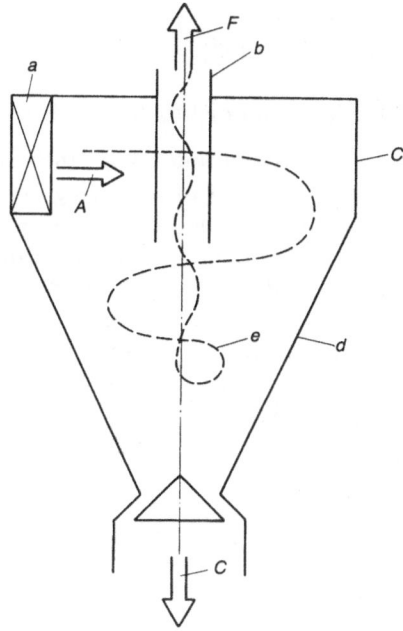

Figure 4.7 Sketch of a cyclone: A, feed; F, fines; C, coarse material; a, inlet; b, outlet pipe; c, top of the cyclone; d, conical section; e, flow path of the air.

size, but which should theoretically remain suspended in circular paths, being dragged downwards and thus enabled to reach the wall and get into the coarse material.

The gas cyclone is employed as a dust separator (see p. 34) because of the advantages of highly safe operation, simple construction and low costs. It is used for fine dusts when the grade efficiency required is not too high. Frequently, however, even fine dusts can be readily separated because of their tendency to agglomerate in the cyclone. Coarser dusts above about 5 μm can be satisfactorily separated and particles larger than 30 μm are generally almost completely collected. The separation is facilitated by increasing the dust loading. At high gas temperatures often only the cyclone can be considered as a separator. Furthermore, the cyclone is employed as a separator before devices for removing the finest dusts (cloth filters, electrostatic precipitators and scrubbers).

In the hydrocyclone the suspension is introduced under pressure whereas a gas cyclone often operates under suction. A core of air is thus formed at the axis of symmetry of the hydrocyclone. Determining the cut point is difficult and pilot-scale laboratory experiments must be used. With high throughputs it is more economic, for the same cut point, to use several small hydrocyclones in parallel than to work with one large one. This is

particularly true for separating fines. To obtain sharper separations hydrocyclones are connected in series [95].

(ii) Recirculating air classifier (with distributing plate)

The feed is thrown off a distributing plate into a vertical, generally
rotating, i.e. helically flowing, stream of air giving cross-current sifting. At
the top the fines are drawn off towards the inner zone – often by means of
a deflecting wheel – and hence counter-current sifting takes place. The
coarse material centrifuged out onto the periphery of the classifier is
separated off there, and, as a consequence of the loss in angular momentum, the boundary layer flows inwards (intensified Taylor–Görtler vortex
with the formation of strands of material) and effects a further sifting of
the coarse material. The recirculating air classifier is illustrated in Fig. 4.8.

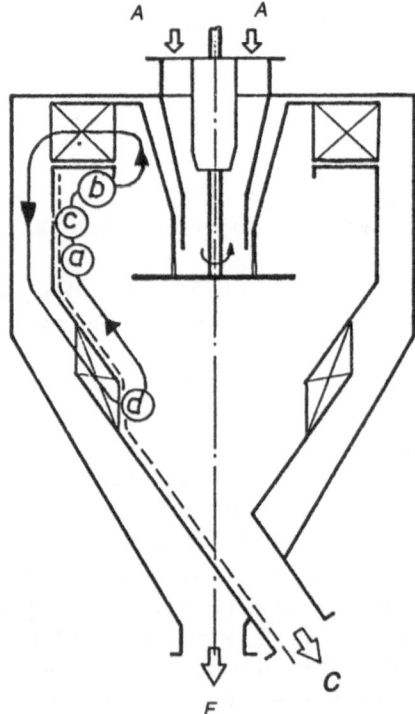

Figure 4.8 Sketch of a conventional recirculating air classifier: A feed; F, fine
material; C, coarse material; a, ascending circulating stream of air (equilibrium
cross-current sifting); b, inwardly diverted stream of air (spiral sifting); c, sifting off
of the coarse material centrifuged out on the wall; d, secondary sifting.

(iii) Disc centrifuge and plate separator

The common disc centrifuge, which has been used for the separation of milk for many years, separates a lighter from a heavier phase in a narrow slit between conical discs. A large number of slits are arranged parallel to one another (Fig. 4.9).

The disperse phase can be lighter (fat globules) or heavier (solids) than the continuous phase. In theory heavier particles move outwards. However, they can be carried away counter-currently by the inwardly flowing liquid. In this case the radial component of their velocity is less than that of the liquid. As a consequence they move in transverse flow onto the opposite disc and from there are carried away and discharged by a superimposed outwardly directed density convection current. The solids are removed from the periphery either intermittently by automatically controlled valves or continuously through nozzles.

(d) Inertial deflecting separation

Deflecting separation is a modified cross-current separation whereby the inertial forces that separate the particles from the current arise through its deflection. When any fluid laden with solids flows around a body, deflecting separation takes place. In engineering practice the principle is applied most frequently in deep-bed filtration using fibres or particles as

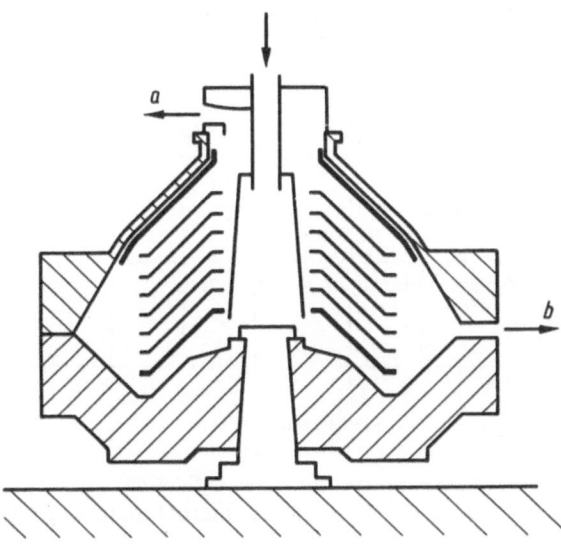

Figure 4.9 Schematic construction of a disc centrifuge; a, liquid; b, solid or heavy liquid.

the collecting medium and in wet separators involving flow around droplets. Deflecting separation also occurs, for example, in pin mills. Here it limits the extent of the pulverizing because very fine particles no longer come into contact with the grinding pins. For a deflecting separator the degree of separation is defined as the ratio of the cross-section of the stream out of which particles still reach the surface of the body around which the stream is flowing to the cross-section of this same body. This index of the extent of separation can be estimated.†

Examples of plots of curves estimated on this basis are shown in Fig. 4.10 for spheres on circular cylinders. The degree of separation depends on the inertial parameter ψ and the Reynolds number Re. ψ is defined as the ratio $\rho_p x_p{}^2 v / 18\eta d_c$ where ρ_p is the density of the sphere, η is the viscosity, v is the velocity of approach, x_p and d_c the diameters of the particle and the cylinder respectively, and $\mathrm{Re} = v d_c \rho_f / \eta$ where ρ_f is the density of the fluid.

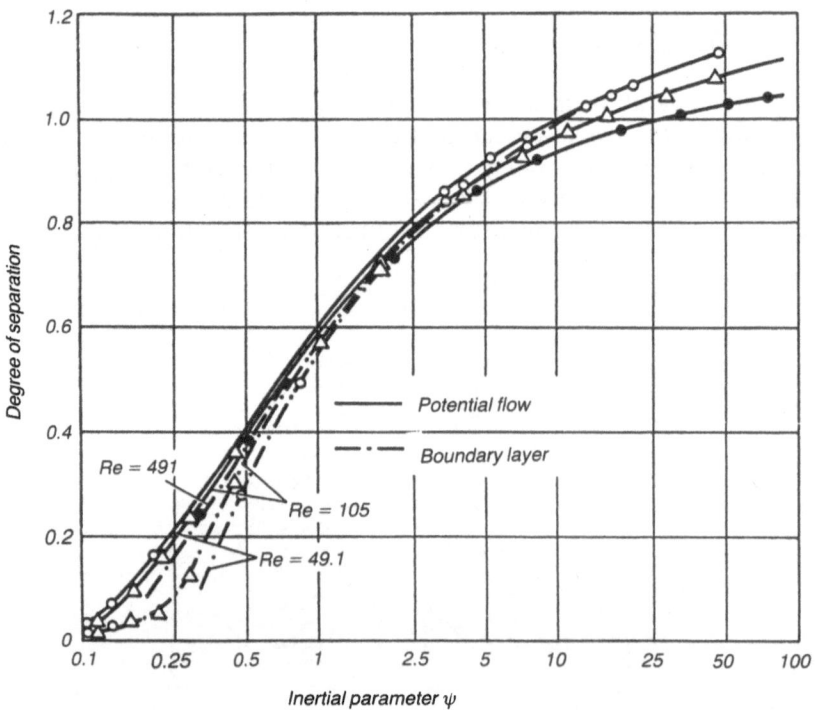

Figure 4.10 Degree of separation for the deflecting separation of spheres by cylindrical collectors as a function of the inertial parameter and the Reynolds number Re [66].

† Translator's note: this measure of the degree of separation (*Trenngrad*) differs from the 'grade efficiency' as usually defined in English-language texts.

The curves plotted are valid for $\rho_p = 1 \text{ g/cm}^3$. The solid curves were calculated for potential flow and the chain curves allow for the effect of the boundary layer.

If particles are to be effectively removed they must adhere to the surface. The overall extent of the separation achieved is the product of the index of separation, as defined above, and the probability of adhesion. The probability that a particle will adhere decreases as its size and velocity increase. Figure 4.11 shows examples of how this depends on the velocity.

The many designs of fibre filters and wet separators will not be described in this survey. Moreover, it is essential to understand that, when very fine particles are removed, separation by diffusion becomes an important transport mechanism in addition to that due to inertia.

(e) Separation in pulsating flow

When the Reynolds number is above the range in which Stokes law can be used, so that a non-linear resistance law applies, coarse particles separate according to their settling velocity under pulsating flow. This effect is made use of in the jigging process. Here, of course, the particle concentration is so great that buoyant separation also occurs as a result of the different particle densities, as is the case in a fluidized bed. Both air and wet jigging are used in mineral dressing, e.g. for the sorting of coal.

4.1.2 Sink-and-float sorting

In order to sort particles according to their density the carrier liquid must have a density between those of the fractions to be separated. This is

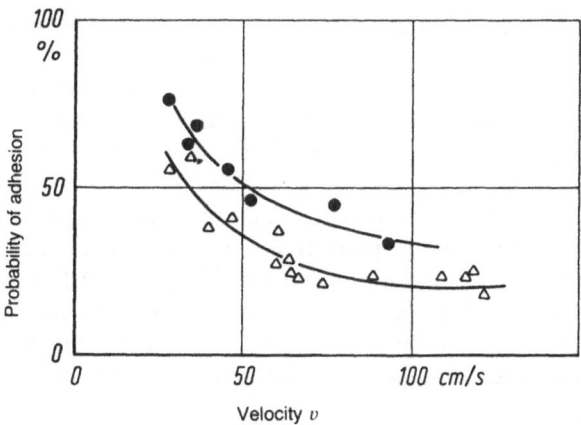

Figure 4.11 Probability of adhesion of 5.6 μm quartz particles on impact with fibres as a function of the velocity v of approach.

achieved in mineral dressing in two different ways. In heavy-slurry sorting (cf. ref. 29, p. 40 *et seq.*) an aqueous suspension containing a suitable concentration of a fine-grained dense material, e.g. pyrites, is used. In the case of flotation (cf. ref. 29, p. 53 *et seq.*) air bubbles are deposited on particles of one type. The air bubbles can be introduced into the slurry in the following way: the air is liberated from a supersaturated solution, or it issues from porous materials or capillaries, or the air and the slurry are intensively mixed in an injection nozzle. The particles attached to the air bubbles float on the surface of the slurry. In order to achieve a selective deposition of the air bubbles it is necessary that the particles be wetted differentially since air adheres only to non-wetted particles. Selective wetting is achieved by the addition of collectors. Normally the bubbles that reach the surface of the slurry burst, causing the particles floating on top to sink again. The air bubbles are therefore stabilized on the surface by the addition of frothing agents. A froth is formed which is skimmed off together with the separated particles. The frothing agents also influence the size of the air bubbles.

Sink-and-float processes generally function hydrodynamically according to the cross-current principle, and often use reaction chambers in series (flotation cells).

4.1.3 Electrical and magnetic separation processes

(a) Electrical separation processes

In electrical separation processes the effect of the force acting on electrical charges in an electric field is utilized. The principle is used in electrical sorting (cf. ref. 29, p. 52 *et seq.*) if the components of a body of materials (e.g. heavy mineral sands) can be charged differently. The sorting is generally done using roll separators. However, in comparison with other sorting processes electrical sorting is limited to a few special applications. The main application is the removal of fine dusts or droplets in an electrostatic precipitator.

Figure 4.12 illustrates the construction of a tube-type precipitator. In addition to this there is a plate-type precipitator in which many flat collecting electrodes are arranged parallel to one another with discharge wires between them. The basic mode of operation is the same for both designs: the particles to be separated are charged electrically and are collected on the tubes or plates to which they migrate cross-current to the stream of gas.

Because of the high d.c. potential at the discharge electrode (which in practice is usually negative) a very strong electric field is produced which, because of the small radius of curvature of the surface of this electrode,

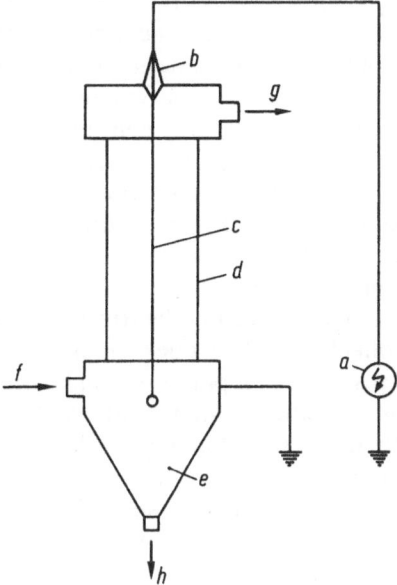

Figure 4.12 Basic design of a tube-type electrostatic precipitator: a, high voltage generator; b, insulator; c, discharge electrode; d, tubular collecting electrodes; e, dust hopper; f, dusty gas; g, clean gas; h, dust.

can assume very high values (8×10^4 up to 5×10^5 V/cm). In this case electrons already present in the vicinity of the discharge electrode are rapidly accelerated away from it. When they collide with gas molecules more electrons are produced and the gas molecules become positively charged. These positive ions are accelerated towards the discharge electrode, hit it and thereby knock more electrons out of it. This process is associated with the emission of light (corona discharge). Further away from the discharge electrode slower electrons accumulate on gas molecules, thereby producing negative ions. At 760 torr there are almost no free electrons at distances of more than 1 cm from the discharge electrode.

The ions present in the electric field of the precipitator charge the particles. For particles with diameters over about 0.5 μm the amount of charge taken up is proportional to their surface area and the field strength. In this size range charging by ionic bombardment predominates, whereas with smaller particles charging is by diffusion. In the latter case the charge is roughly proportional to the particle diameter.

The migration velocity of negatively charged particles towards the collecting electrode is determined by the equilibrium between the electric field force and the drag force. In the range of charging by ionic bombardment this velocity is proportional to the size of the particles but in the range of ionic diffusion it is independent of their size. This explains the good separation that is achieved for very small particles. If the migration velocity of the particles is known, the collection efficiency can be calculated. However, in some instances the experimentally determined migration velocities deviate substantially from the theoretical values. Obviously the geometric configuration of the collecting electrodes also exerts an influence. Therefore, prior estimation of the size of an electrostatic precipitator is based on experience.

For good precipitation the specific electrical resistance of the dust layer formed on the collecting electrodes must lie between 10^4 and 2×10^{10} Ω cm. This resistance depends strongly on the condition of the gas. With dusts of high specific resistance (above 2×10^{10} Ω cm) the charge does not escape, with the result that high field strengths are produced in the pores of the dust layer. Breakthroughs (flashbacks), which upset the precipitation, occur at inhomogeneous spots in the deposit. Under some circumstances this can be remedied by injecting water (irrigated electrostatic precipitation).

The collecting electrodes are cleaned by rapping them at regular intervals so that the dust layer slides down into the dust hopper. This creates a cloud of dust whose particles enter the stream of gas again. It is expedient to divide the length of the precipitator into sections.

Electrostatic precipitators have the disadvantage of high investment costs and they suffer from the restriction that the dust to be removed must have certain electrical properties (specific resistance between 10^4 and 2×10^{10} Ω cm). However, since they easily remove very fine particles (less than 1 μm), there is often no alternative to installing such a device. Their advantages are that they cause small pressure losses, their energy costs are low and they are cheap to maintain. Moreover, they can be operated at temperatures up to about 450 °C.

(b) Magnetic separation processes

In magnetic separation the materials are sorted on the basis of their differing behaviour in a magnetic field. The particles are separated according to their permeability. Weakly magnetic materials can be separated only by the dry magnetic procedure which is applied to particles of size below about 1–3 mm. However, with strongly magnetic materials (e.g. magnetite and ilmenite) particles as large as 20–30 cm can also be separated. Here the wet magnetic procedure is favoured. The usual kinds of magnetic separators are belt, roll and drum separators of many different designs (cf. ref. 29, p. 47 et seq.). The most versatile are the drum separators. They can be employed for all particle sizes and are operated under both dry and wet conditions. The main applications of magnetic separation are the beneficiation of iron ores and the sorting of heavy mineral sands. The procedure also plays an important role in the removal of iron from the raw materials of the glass and ceramic industries, and in ridding bulk solids or suspensions of troublesome iron parts or tramp iron that would damage apparatus downstream.

4.1.4 Separation processes using wall friction

Separation processes using wall friction are related to those involving separation in fluid streams. However, the separation occurs not as a result of the relative motion of freely suspended particles in the fluid but as a result of movement on a table. The drag force is replaced by wall friction. The drag force near the wall is not eliminated but is of little consequence for the separating operation. Much more relevant is the differing friction between the particles and the wall which is caused by the variation in their surface roughness, shape and hardness. The inertial effects which apply are due to gravity and to the reciprocating motion or vibration and generally to a combination of all these forces. The methods are used in mineral dressing and food technology.

A typical example of this separating principle is the shaking table on which the slurry is fed at the upper end onto an inclined plate, usually filled with grooves, that is moved to and fro. The slurry spreads over this plate, which is known as the table, and the particles with a large weight-to-friction ratio arrive at the lower edge more quickly and in a different position from those experiencing more friction.

A second example is the paddy sorting table for the separation of husked and unhusked rice or oats. In this device an oscillating inclined table is also used. It is equipped with triangular solid bodies that bounce around and transfer an upwardly directed impulse to the mass of cereal grains sliding hither and thither over the table. By this means the smooth unhusked grains migrate upwards and, under the influence of their greater weight, the raw husked grains covered with a fine hairy fluff migrate downwards. As with all sorting processes the grains must be carefully graded beforehand.

4.1.5 Sifting

The sifting procedure is used to separate an assembly of particles into two size grades by means of a sieve. For grading into several fractions two or more sieves are stacked on top of one another. The cut size is determined essentially by the width of the sieve apertures. However, the particle size distribution and the amount of the feed, the movement of the sieve and the sifting time also play an important role.

For sifting the effective forces are gravity, hydrodynamic and aerodynamic forces, impact forces, frictional forces and, particularly in the case of fine particles, adhesive forces. For undersize particles the probability of passing through a sieve aperture is not necessarily equal to unity. The nearer the particle's size is to the mesh width, the greater are the obstacles that such particles must overcome in passing through the mesh and the greater is the danger that they will stick in the mesh. If the ratio of particle size to mesh width lies between about 0.8 and 1.0 there is a greater danger of blinding the apertures. For values below 0.8 and greater than about 1.5 the danger is slight. If the feed contains large particles the sifting is expedited since, to some extent, such particles release trapped granules from the mesh. As the amount of feed is increased, the migration towards the mesh of the fines within the heap of material moving around on the sieve cloth or plate is hindered and the sifting time needed to achieve a sufficient sharpness of cut is increased. In general a longer sifting time improves the sharpness of cut and raises the cut point.

The sifting procedures used in practice are generally independent of the particle size in the coarser range. Only in the fine range, particularly below about 50 μm where the adhesive forces increase substantially in relation to the other forces acting on the particle, must special procedures be adopted. The adhesive forces are largest in moist material (see section 3.6). Compared with the hydrodynamic forces they are small in the case of wet sifting in which the particles suspended in a liquid are rinsed through the apertures. Wet sifting is therefore most suited to fine material. If it cannot be applied, sonic sifting or air-jet sifting is used. With sonic sifting not only is the frame of the sieve set vibrating – as with the usual vibratory sieves – but also the sieve cloth is directly activated to vibrate in many places by means of agitating tappets so that it experiences accelerations up to 15 times the acceleration due to gravity. In the case of air-jet sifting (particle size range 40–100 μm) a drum covered with a sieve cloth (Fig. 4.13), in which the feed is rolled around, is flushed from outside with a sharp jet of air so that blinded apertures are continually blown free and the agglomerates that form are broken up. The circulating air flows from within the drum outwards through the sieve and takes the fines with it. They must subsequently be removed from the air in a separator.

Vibratory sieving machines can be operated either mechanically (eccentric and out-of-balance drive) or magnetically. The sieve frame can be positively agitated (eccentric drive) or allowed to swing freely (out-of-balance and magnetic drive). With mechanical drive the throw of vibratory sieving machines is about 15 mm and the frequency is about 10–20 Hz. Sieves with grates or punched plates are used for sizing coarse material, sieves made of woven wire mesh are used for fine material and sifting surfaces of rubber or plastic are used for sticky or very abradable material.

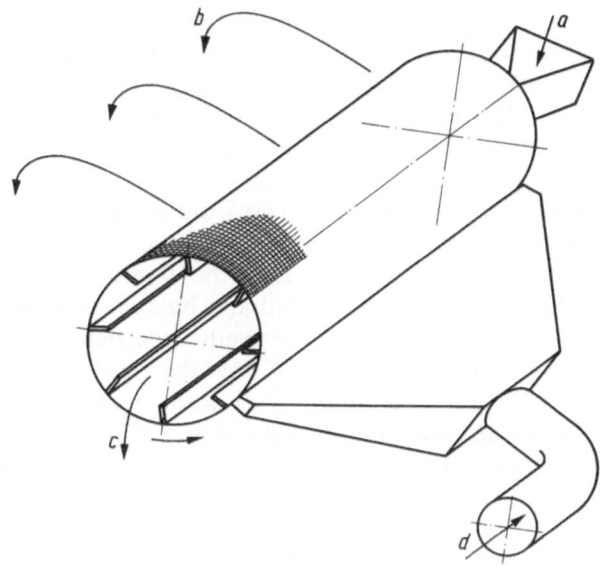

Figure 4.13 Air-jet sieve: a, feed; b, fine material and air; c, removal of coarse material; d, air.

4.1.6 Cake filtration

The separation of solids in filters or sieve centrifuges is included among the cake filtration processes. A solid cake is formed on the filter medium (cloth, porous ceramic, wire mesh, sieve plate etc.) and particles entrained in the stream of liquid are separated out in the cake.

To fix the size of the filtering area of a filter or a sieve centrifuge it is necessary to know how the thickness of the cake and the amount of filtrate will increase with time. The filtration process can be designed if some simplifying assumptions are made. The most important of these are as follows: laminar flow through the cake, and a cake whose porosity and structure do not change with respect to time and location (homogeneous incompressible cake). The duration of the filtration is ascertained by carrying out a materials balance for the solid phase (the amount of solids flowing in during the time t is equal to the amount of solids in the cake), and, in addition, Darcy's filtration equation (eqn (3.51)) is used to describe the passage of the liquid through the cake and the filter medium. A differential equation is obtained for the volume of the filtrate as a function of time (the filter equation). In principle this can be integrated if the dependence of the pressure drop on the time or the volumetric flow rate (characteristic curve for the pump) is known. In a filter the pressure drop is produced by a pump either as excess pressure or partial vacuum, but in a centrifuge it is produced by the annulus of slurry rotating over the cake. The other quantities in the filter equation depend to some extent on the structure of the filter cake. However, this is not amenable to a purely theoretical treatment, and therefore these quantities have to be determined by means of an experimental test using a small-scale filter plate.

It has been observed that in many cases the filter cake is compressed to a greater or lesser extent under the load of the effective pressure drop. This manifests itself particularly clearly in the case of fine-grained or easily deformable solids, whereas coarse, rigid and bulky solids are more inclined to form incompressible filter cakes. To predict the behaviour of compressible filter cakes it is necessary to know how the local porosity depends on the pressure, under the assumption that, with increasing layer thickness, the changes in the porosity of the cake will behave similarly. A similar cake structure does not exist if, in the course of time, the pores are blocked by fine particles (internal filtration) or by gas bubbles (vacuum filtration).

The sequence of operations during filtration is as follows: cake formation, removal of residual moisture, washing (more or less complete rinsing away of the filtrate), removal of residual moisture, removal of the cake and flushing (cleaning of the filter medium). All these operations can, in principle, be carried out not only in ordinary filters but also in sieve centrifuges. The main criteria for choosing filters or sieve centrifuges are throughput, residual moisture, purity of the filtrate and running costs. Sieve centrifuges are generally to be preferred if a particularly low residual moisture content is absolutely necessary. For fine-grained material filter media with correspondingly fine pores must be used. If the rate of filtration is slow, the filtering area must be correspondingly large. Therefore filters are more often used for large volumetric flow rates. They can function either continuously or batchwise.

Continuous, i.e. rotary, drum filters usually operate under vacuum. The maximum pressure difference across the filter is thereby limited to about 70 kPa, and this also limits the extent to which the residual moisture content is reduced. Very fine slurries with extremely compressible cakes can be filtered better in drum filters if the structure of the cake is improved by means of filter aids. These can either be added to the pulp or be applied to the filter as a pre-coat layer before the pulp is introduced. Other continuous rotary filters are disc and cross-flow filters. With slurries that contain rapidly settling constituents, sedimentation must be prevented either by stirring (as in the case of the drum filter) or by arranging for the stream to approach the filter in the direction of sedimentation. The latter happens in the cross-flow and continuous-belt filters for example (Fig. 4.14). If it is required to increase the extent to which the cake is dried, pressure filtration often remains as the only solution; at medium

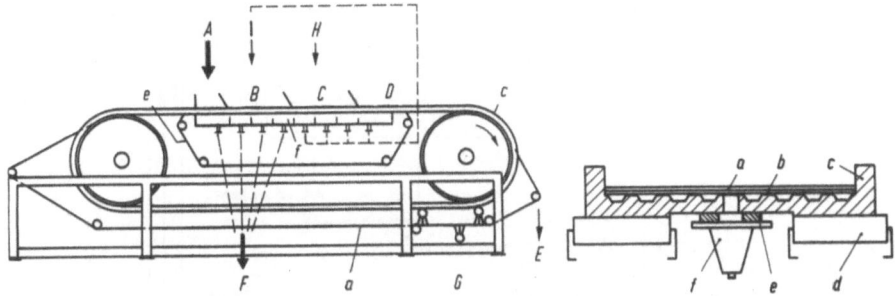

Figure 4.14 (a) Belt filter with counter-current washing (Lurgi); (b) cross-section through the belts: A, feed; B, first wash; C, second wash; D, drying zone; E, cake discharge; F, filtrate; G, belt wash; H, water; I, washing filtrate; a, filter cloth; b, perforated supporting rubber belt; c, travelling belt; d, guiding belt; f, suction box.

pressures the continuous rotating-leaf pressure filter can be used, and a large number of designs for higher pressures (up to about 3 MPa) are available using batchwise or automatically controlled cyclic operation. They differ from one another in the manner in which the cake is removed and transported. Typical representatives are filter presses and leaf and cartridge filters.

For smaller throughputs involving fine-grained materials discontinuous sieve centrifuges are also used, particularly when low residual moisture contents are required. Continuously working sieve centrifuges are suitable mainly for coarse crystalline materials because only smooth sieves can be used on account of transport properties and attrition. High throughputs are attainable with these separators. The transport of the cake either takes place continuously as a result of the centrifugal force or an additional motion (conical drum, oscillating centrifuge, scroll-type centrifuge, tumbling centrifuge) or is carried out intermittently by means of a reciprocating plunger (pusher centrifuge). The throughput of the filtrate depends on the level of the liquid. Figure 4.15 shows the basic design of a pusher centrifuge. A new development (Krauss–Maffei) increases the throughput considerably during the stage when the residual moisture is being removed; it does this by means of a device appended to the sieve that acts like a syphon on the level of the liquid flowing through.

4.2 MIXING PROCESSES

4.2.1 Mixing of gases

Two or more gases will mix spontaneously and completely by diffusion after a sufficiently long time. The rate of approach to a uniform concentration in such a system can be increased by vigorous mixing of the

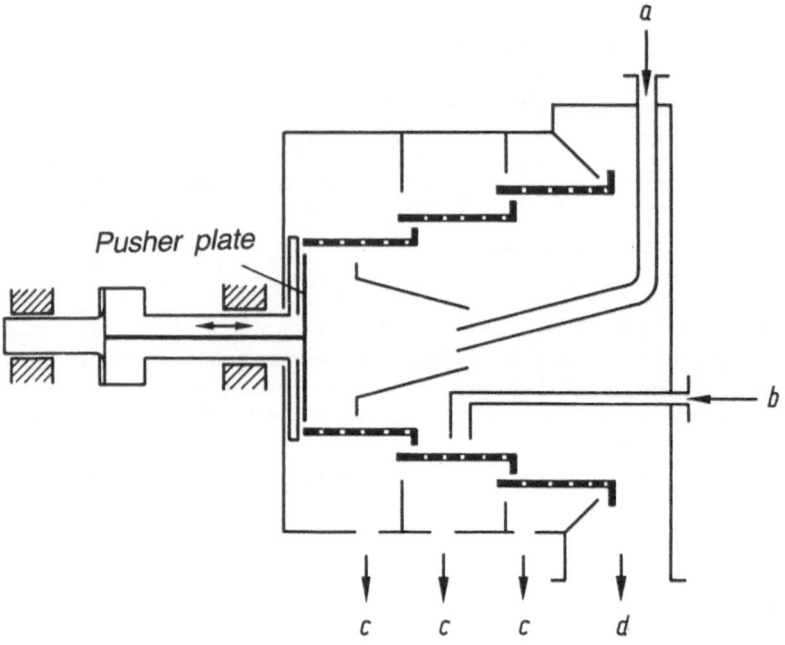

Figure 4.15 Pusher centrifuge: a, slurry feed; b, washing liquid; c, filtrate; d, solids.

components through convection and turbulence because in this way the average size of the domains occupied by the individual components of the mixture is reduced.

Turbulent mixing is effected, for example, with the aid of a free jet of one component which is blown into the other quiescent component (air conditioning). Often the mixing effect of the free jet is part of a complicated mixing process in a mixing device. Examples of such devices are mixing tubes, with or without inserts (bends, shutters etc.), injectors or mixing chambers of various designs.

4.2.2 Mixing of a gas into a liquid (aeration)

Aeration is used in a wide variety of chemical reactions to increase the interfacial area between the gaseous and liquid phases. In the range of lower viscosities stirrers are employed, and kneading machines are used for highly viscous liquids. Two types of stirrer are available: hollow stirrers which suck the gas in as they rotate and distribute it throughout the liquid (Fig. 4.16), and radially transporting fan turbine mixers or disc agitators

Section 4.2 was written with the cooperation of Dr H. Reichert.

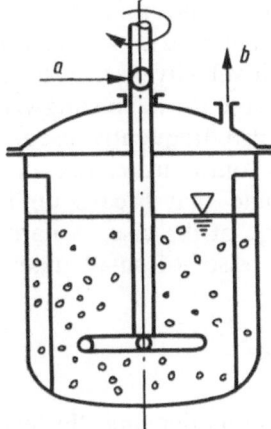

Figure 4.16 Diagram of a hollow stirrer for aerating waste water: a, incoming air; b, outgoing air.

into which the gas is introduced separately from below, through, for example, a perforated rim. The simplest and most effective hollow agitators are tubular agitators. Because of their limited sucking capacity hollow agitators are not suitable for very high gas throughputs (Fig. 4.16).

4.2.3 Dispersing of a liquid into a gas (nebulization)

Liquid phases (liquids, suspensions) are dispersed in a gas by means of nozzles and nebulizers. A distinction is made between single-component jets (spiral, slit-shaped and impact jets) and two-component jets. As an example of a single-component jet Fig. 4.17 shows the construction of a spiral-chamber nozzle in which the liquid flows in tangentially. All single-

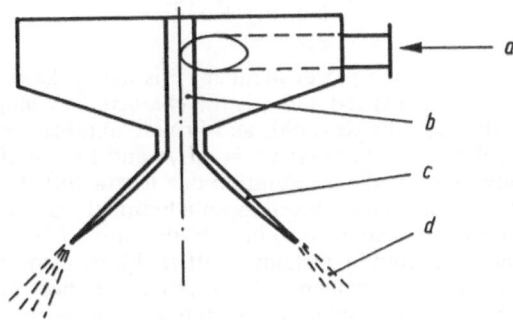

Figure 4.17 Principle of a spiral-chamber jet: a, liquid; b, core of air; c, liquid film; d, drops.

component jets first produce a thin film of liquid that eventually breaks up into drops. In two-component nozzles a stream of liquid, concentric with a ring-shaped stream of high-velocity gas, is torn apart by this gas stream. These nozzles are also suitable for nebulizing viscous media.

Rotary nebulizers are also frequently used. In the simplest case they consist of a spinning disc onto the centre of which the liquid phase is introduced. A film of liquid that breaks up into threads and then into drops is formed on the edge of the disc. A narrow range of droplet sizes is achieved by working with discs of small diameter, high rotational speed and low liquid throughput.

4.2.4 Mixing of liquids

The mixing operation is thermodynamically favourable for soluble liquids. As in the case of mixing of gases, the function of the mixing equipment is to make the constituent domains of the components as small as possible in order that the diffusive levelling out of concentration can be achieved more rapidly. When liquids which are not soluble in one another are mixed (homogenization, emulsification) the function is the same. Different mixing equipment is used depending on the viscosity of the liquids.

For liquids of low and medium viscosity agitators with various configurations are used. They can be classified according to the principal flow patterns produced (tangential, axial and radial agitators) and the range of viscosity. For thick liquids slowly running anchor and helical agitators are used. Whirling of the liquid in the mixing vessel is prevented by the installation of baffles or, in the case of gentle stirring, by mounting the stirrer eccentrically.

The following relation, which is obtained by dimensional analysis under the assumption of geometrical similarity, is used to calculate the power needed to drive the rotor of an agitator:

$$\text{Ne} = f(\text{Re}, \text{Fr}) \tag{4.1}$$

where $\text{Ne} = P/(\rho n^3 d^5)$ is the power number (P is the brake power, ρ is the density, n is the rotational speed and d is the characteristic length, e.g. rotor diameter). Re and Fr are the Reynolds and Froude numbers respectively, which are usually defined as follows: $\text{Re} = nd^2/v$ and $\text{Fr} = n^2d/g$ (v is the kinematic viscosity and g is the acceleration due to gravity). If the height H of the liquid in the vessel is introduced as an additional variable, then the above expression must be extended to include the ratio H/d. The formation of a vortex can be prevented by installing baffles. Many measurements which can be used to obtain the dependence of the power number[†] on the Reynolds and Froude numbers are available for common agitator designs. For example, Fig. 4.18 shows plots of $\text{Ne} = f(\text{Re})$ for the case of very high Froude numbers, i.e. when no vortex is formed. The laminar region for $\text{Re} \lesssim 5$ and the turbulent region for $\text{Re} > 10^4 - 10^5$ can be seen.

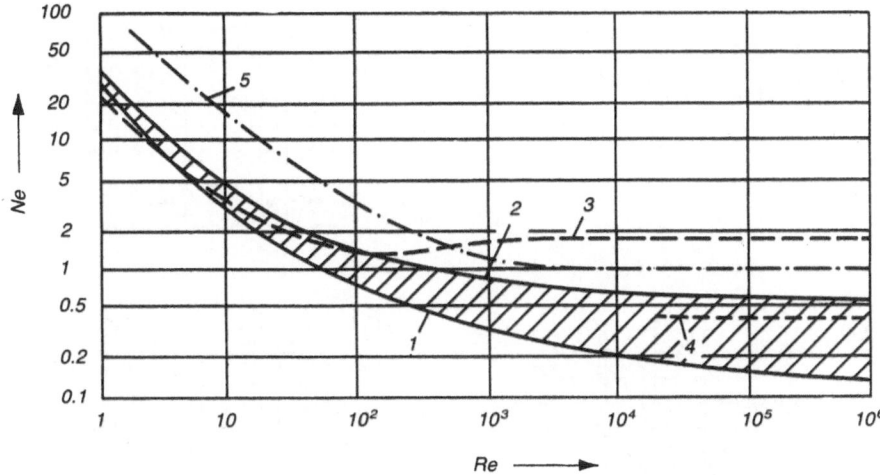

Figure 4.18 Power number as a function of Reynolds number for various types of agitator (after F. Kneule): 1, propeller agitator without baffles; 2, propeller agitator with baffles; 3, paddle agitator with baffles; 4, paddle agitator without baffles; 5, turbine agitator with shroud.

Soluble liquids are often mixed in pipelines. To improve the mixing effect of the turbulent flow in the pipeline, one component can be introduced through an injector or a venturi nozzle. Viscous liquids can also be mixed in pipelines by means of mixing devices without moving parts. Such devices consist of inserts in the pipe that repeatedly and successively split up and displace the stream of liquid.

Figure 4.19 shows schematically how a mixing process involving the laminar flow of highly viscous liquids proceeds. In this case mixing by diffusion does not occur. An arrangement of thick parallel layers of the components to be mixed is divided up (shaded areas) and subsequently the newly formed layers are mutually displaced (arrows). Division into thinner layers is also achieved by shearing onto which a transverse displacement must then be superimposed. Laminar mixing is mechanically (not thermo-dynamically) reversible, as long as the total mass sticks together, with the result that it can always be brought back into the old state by reversing the direction of motion. If, as in a kneader, portions are removed from the mass, and the mass is thus divided and recombined, irreversible mixing occurs. By repeating the processes of shearing and displacement or division

† Translator's note: Rumpf calls the power number the Newton number which, in turn, is the same as the drag coefficient. It is possible to show that these dimensionless groups are essentially the same. See, for instance, W. L. McCabe and J. C. Smith, *Unit Operations of Chemical Engineering*, 2nd edn, 1967, pp. 253–6, McGraw-Hill.

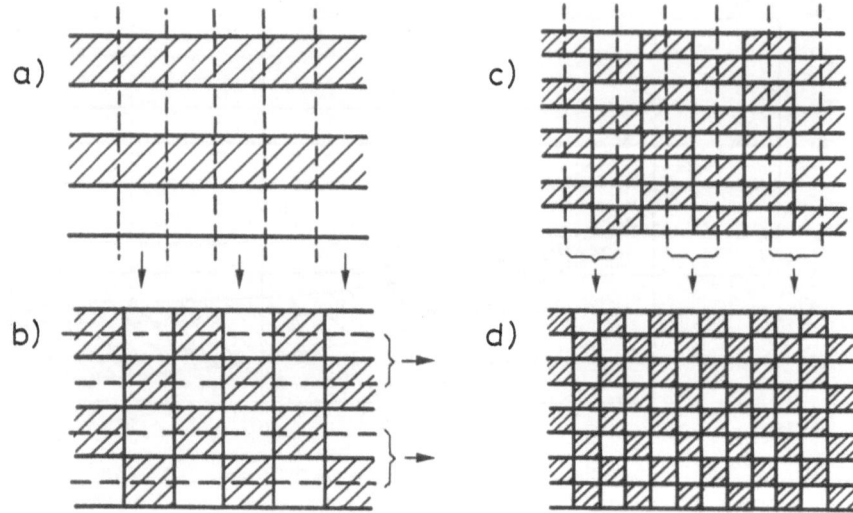

Figure 4.19 Basis of mixing by division and displacement.

and recombination, the quality of the mixture can theoretically be improved as much as desired. This principle is used in mixing in kneaders and internal mixers, where the shear flow between the kneading blades and the wall of the housing makes an important contribution to the mixing effect.

4.2.5 Dispersing of solids in gases

The problem of dispersing solids in gases occurs mainly in the case of gas–solid reactions. In the range of high solids concentrations the mixing is carried out in a fluidized bed.

Most fluidized bed reactors function in the range of aggregative fluidization. With a further increase in the superficial velocity the bed changes into a condition whereby particles are discharged from it. The rising fluidized layer has the character of a cloud of fine dust and a lower concentration of solids than the bed from which particles have not been removed. The discharged solids must be continuously replaced by newly added material or returned to the reactor. By keeping a steady flow of gas and a uniform loading of solids, stable clouds of fine dust can be maintained from which few solids are discharged. Clouds of fine dust can also be produced in a reactor that is designed as a diffuser (Fig. 4.20). Here the solids are continuously discharged from the narrow part and subsequently fall down again in the wide cross-section of the diffuser.

A gas–solid mixture with a low solids content can also be obtained by

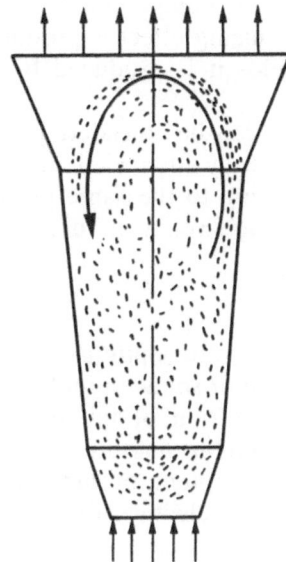

Figure 4.20 Fine-dust reactor (venturi reactor).

jetting the solid into a mixing chamber. This procedure is used in the case of pulverized-fuel firing (cyclone firing) for example. The solid can be de-agglomerated before the actual mixing by introducing it as a jet or by hurling it in. This factor can play an important role, particularly in the case of very fine dusts.

4.2.6 Dispersing of solids in liquids

The mixing of solids with liquids to form suspensions is carried out, depending on the viscosity of the liquid, in agitators or kneaders. When the particles are very small de-agglomeration often has to be achieved here as well. All mixing devices distribute and de-agglomerate. De-agglomeration can take place by the mechanical action of the mixing implements, by the interaction of the particles or because of the hydro-dynamic forces. Not only de-agglomeration but also the prevention of re-agglomeration can be assisted by the action of surface-active additives (see section 3.7.2). The action of the dispersants is due to two factors: they are adsorbed on the surface of the solid, and, depending on the nature of their chemical composition, they effect either a spatial separation or an electrostatic repulsion of the particles.

More generally, all de-agglomeration processes in a liquid phase can be considered as solid–liquid mixing processes. These include, for instance, de-agglomeration by means of ball mills, stirred ball mills, colloid mills and roll stands. All these types of equipment de-agglomerate and mix. With liquids of low viscosity the mixing effect is produced mainly by turbulence, but with high-viscosity liquids it is produced by shear flow and displacement (see above).

Mixing in a fluidized bed can be considered as a further mixing process involving solids that are already immersed in liquids; this mixing can be carried out in liquids according to the same principles as those that apply in the case of gas–solid mixing already discussed.

4.2.7 Mixing of solids

A uniform random mixture is the optimum state of mixing that can be achieved by the bulk handling of solids in industrial mixers. In this case the expected value of the composition of the mixture is the same at every point within it. A uniform random mixture can only be attained if there is no favoured direction of movement of individual solid particles by the mixing implement and if no selective forces come into play; if these conditions are not satisfied, demixing occurs. The attainment of a uniform random mixture depends to a considerable degree on the size, density and shape of the particles. In drum mixers, for instance, not only small particles but also denser ones tend to accumulate at the bottom of the mixer. Very fine particles can agglomerate.

The methods for mixing solids can be classified as follows, along with the processes that occur in the usual mixing devices:

1. mixing in an aggregate (bulk solid)
2. mixing by hurling and spinning (a) under gravity and (b) in a centrifugal field
3. mixing in a suspended state
4. mixing in a fluidized bed
5. mixing of streams of material

When bulk solids are mixed a motion is introduced within the body of material such that the particles are displaced relative to one another. The motion of the mixing elements must be slow if only this mechanism is to take place (e.g. slowly running screw mixers). In this way steady state flow is established in the bulk material as it flows around the mixing device. At higher speeds of the mixing implements the particles are thrown out of the body of material or air can be drawn in. When this happens centrifugal mixing also occurs or the bulk mixing becomes fluidized-bed mixing.

When centrifugal mixing occurs portions of material are thrown out of the bulk and, after being rearranged in the free space of the mixer, they fall back again onto the surface of the bulk of the material. Gravitational or centrifugal forces can play a decisive role in controlling the surface area of the body of material and in returning the separated material to it.

In the case of mixing in a state of suspension the material is completely suspended in a gas. The suspension can be maintained either by a sufficiently fast stream of gas, as in pneumatic conveying, or by rapidly moving mixing implements.

If the solid components to be mixed flow together in uniformly added streams of material they mix themselves at the point where they merge. These streams can be flowing bulk solids (e.g. when a mixing silo is being emptied) or they can be pneumatically transported or fluidized streams. The mixing process yields a random mixture only when a random operation is superimposed on the transport process as is the case, for example, in a fluidized bed or in the dispersion of the streams of material at the point where they merge.

4.3 AGGLOMERATING AND COAGULATING PROCESSES

During the production and processing of powders it is observed that adhesive phenomena occur to an ever-increasing extent as the grains become finer. Agglomeration can occur as a result of collisions between the particles when they are dispersed in a liquid or a gas. This takes place in bulk solids when agglomerating conditions prevail, which can be the case in dry as well as moist material. In this regard a fundamental distinction between two kinds of processes must be made.

In **cumulative granulation** the particles deposit themselves spontaneously on one another, as is the case in a mixer, a nodulizing drum, a disc pelletizer, a sieve or a fluidized bed.

With **pressing processes** the material is forced to form agglomerates by being compressed. The amounts of material formed into tablets can be apportioned before they are pressed; in the case of compaction in a roller mill these amounts are determined by the nip condition of the mill. With extrusion presses the feed is introduced either intermittently or continuously and the extruded material is broken up into agglomerates at the mouth of the press.

Adhesion between particles can be brought about by the presence of liquids, by binders or by the formation of bridges of solid matter, and can be due to van der Waals and electrostatic forces which act regardless of these bridges. Adhesion determined by shape plays a part in conjunction with other adhesive mechanisms in the production of paper, felt and synthetic fleece and in the pressing of scrap material. With ferromagnetic materials magnetic forces of attraction also come into play. Electrostatic forces are of importance as an adhesive mechanism mainly in the case of cumulative processes, and are less effective in pressing processes. All

Section 4.3 was written with the cooperation of Dr H. Reichert.

remaining adhesion mechanisms can be realized in both cumulative and pressing processes. These processes can be carried out in such a way that certain adhesive mechanisms predominate.

4.3.1 Cumulative granulation

Cumulative granulation is also called **pelletizing**, and is defined as the formation of agglomerates, which are usually not very strong, by rolling and mixing movements in a rotating apparatus or by turbulent movements in suspensions of particles which cause interparticle collisions. Except when very fine-grained material is used, these processes require the application of moisture or other binding agents. In many cases the strength finally achieved, or the critical strength, is obtained only through after-treatment of the agglomerates, e.g. drying, sintering etc. Cumulative granulation is used today on a very large scale for the pelletizing of ores and in the preparation of cement and minerals, fertilizers, dyestuffs, chemicals, pharmaceuticals, foodstuffs and fodder. The cohesion of moist or 'green' agglomerates depends essentially on the action of capillary attraction.

Figure 4.21 shows the three most frequently used devices for cumulative granulation: the disc pelletizer, the conical pelletizer and the nodulizing drum. All employ the same cumulative mechanism: they rotate on their axes, and the grains form spheroidal agglomerates as a result of the rolling motion of the material which is moistened with a liquid binder (usually water) that has been sprayed on. After the formation of a nucleus the granules can be enlarged further by means of various mechanisms: coalescence of pellets already formed, accumulation of abraded particles, fracture of the granules and combination of the fragments, and snowballing. Whichever of these mechanisms predominates, whether the growth, for example, is determined more by coalescence or comminution and combination of the fragments, depends on the strength of the agglomerates being

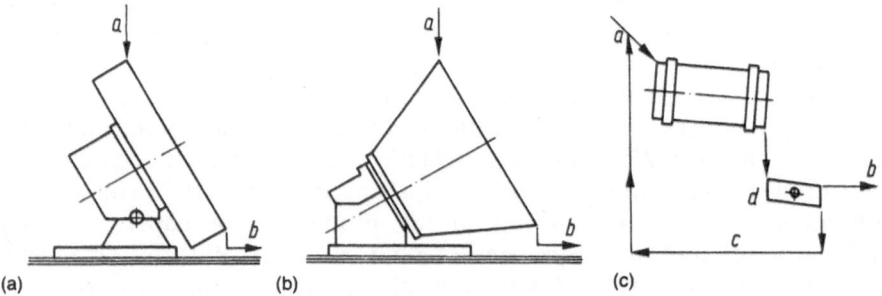

Figure 4.21 Schematic representation of devices used for cumulative granulation: (a) disc; (b) cone; (c) drum; a, feed plus water; b, green agglomerates for further treatment; c, recycle stream (undersize); d, sieve.

formed, i.e. on the fineness of the grains and the addition of moisture. The uniformity of the manner in which the pelletizing equipment functions is very important for its operation and for quality control. Fluctuations in the agglomerating conditions caused by the composition of the feedstock, its charge rate and the charge rate of the water can reinforce themselves and lead to unsteady agglomeration behaviour and fluctuating states of filling. When disc or conical pelletizers are used, the mechanism of the motion of the feed causes a classifying effect. If care is taken to carry out the conglomerating process smoothly, agglomerates with very uniform properties can be produced. In particular, no recycling of the undersize is necessary.

Since green agglomerates have little strength and the forces of adhesion disappear for the most part when the liquid is dried out, the granules must be consolidated by solid binders that become effective during drying (e.g. substances that crystallize out) or by burning (sintering). In the iron-ore industry in particular, whole systems have been developed for producing sintered pellets from fine ores, starting from cumulative granulation on discs, in cones or in drums.

Continuously or intermittently working closed or open mixers with or without mixing implements, as well as vibrating or pulsating chutes or troughs, can also be used for cumulative granulation. The quality of the granules produced is poor; in particular they usually have a rough surface and their size is very non-uniform. Another method of cumulative granulation uses the collision of particles in a turbulent gas fluidized bed to form granules by agglomeration (Fig. 4.22). Powder and binder are introduced from above into or onto the fluidized bed of a venturi fluidizer where agglomeration ensues. If the fluidizing gas is heated, the granules are dried at the same time. As a result of the classifying action in a fluidized bed, new particles of powder only accumulate on the agglomerates until, in counter-current flow to the hot gas, they are carried away downwards through the throat of the nozzle. This is why the size of the granules is very uniform. The procedure, which can start with either powders or slurries (e.g. filter sludges), is particularly suitable for producing fine granules with sizes of the order of 0.75–4 mm.

4.3.2 Agglomeration by compaction

The processes of agglomeration by compaction are characterized by the action of large forces on a particle assembly. For dry material pressures of the order of 10^4–10^5 N/cm^2 are applied or, between rollers, forces of 10^4–10^5 N/cm. With moist material the pressures are much smaller, of the order of 10–10^3 N/cm^2. The pressing of moist masses through the openings of sieves, rollers or dies mainly fulfils the function of impressing shape, i.e.

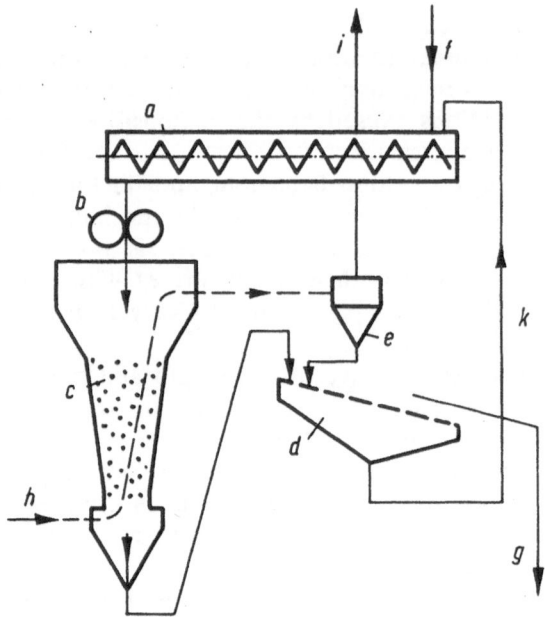

Figure 4.22 Schematic representation of a possible method for cumulative agglomeration in a gas-fluidized bed: a, mixing screw; b, distributing rolls; c, venturi fluidizer; d, sieve; e, exhaust air cyclone; f, feed; g, granular powder; h, (hot) gas; i, exhaust air; k, recycled undersize.

forming. Temperature-sensitive material softens under compressive deformation at the point of contact. Bridges of solid material are then formed. Additional heating encourages plasticizing or partial sintering. The drying of a moist body of material leads to the crystallizing out of dissolved salts and the formation of solid bridges.

If very fine powders or substances that are plastically deformable are pressed, the addition of a binder is not necessary. The cohesion of the pressings is then effected by van der Waals forces, and adhesion is effected by melting and by factors that are determined by the shape of the grains. Substances with low melting points flow together at the grain boundaries to form a homogeneous structure; similar bonds can be produced in almost all materials by hot pressing. Naturally occurring constituents in the material being pressed can, under some circumstances, be effective as additional binders. Dry or moist binders and, perhaps, lubricants need only be used for a few powders which are difficult to compress.

Bench, moulding, stamp, roller and extrusion presses are used to achieve agglomeration by compaction.

In bench, moulding and stamp presses the material to be compacted is pressed in dies by reciprocating stamps or it is formed by reciprocating split moulds. The pressings are distinguished by a very exact shape.

Machines of high capacity have been developed for powder metallurgy and for the preparation of tablets. Whereas in the case of powder metallurgy machines of widely differing designs are used, the **eccentric** and **rotary presses** are well established for the manufacture of tablets.

In eccentric presses only the upper stamp exerts the compression; the lower stamp expels the finished tablet. The disc remains stationary while the feeding hopper and the shoe containing the material to be pressed move backwards and forwards. Filling the die, compaction and expulsion follow in succession while the cam makes a complete revolution. In a rotating press the compacting pressure is exerted by both the upper and lower stamps. The feeding hopper and shoe are stationary. A large number of dies with the corresponding stamps are arranged on a rotating disc, which is known as the circular table. As the disc rotates the holes in the dies pass in succession under the feeding shoe and the devices that activate the stamp to press and eject.

Roller presses for briquetting and compacting use two rollers of equal size that rotate at the same speed in opposite directions and have either smooth or profiled surfaces. The material to be processed is fed above the rollers, drawn in, and shaped and compacted as it passes through the machine. Rollers whose surfaces have cavities that determine the shape of the product are preferred for briquetting; compacting machines use smooth, corrugated or continuously profiled rollers for producing smooth or profiled plates, strips or pressings. Figure 4.23 shows the construction of a roller press for briquetting.

It is often required that the size of the agglomerates in a granulated product should only vary within narrow limits around the average value, that their shape should not deviate too much from what is desired and that their strength should be sufficient to withstand any stresses on them that might occur later. Under these conditions the compacting–granulating procedure is frequently the most economic solution. In this case the powder to be granulated is first compacted into spalls or large briquettes and then crushed to form grains.

Extrusion presses use the resistance that results when a plastic mass flows through channels or open-ended dies to shape and compress material. Binders are usually necessary to produce bonds. Only materials that melt easily or plasticize can be processed without a binder.

4.3.3 Sintering

Sintering is the caking of a material at temperatures in excess of about 50% of its absolute melting point, so that it does not melt during the process. The steps involved in this process are the growth of necks, the rounding off and formation of closed pores and final compaction as the pores migrate. The driving forces are the surface tension and/or the vacant-site concentration gradients related to it, and the effect of curvature

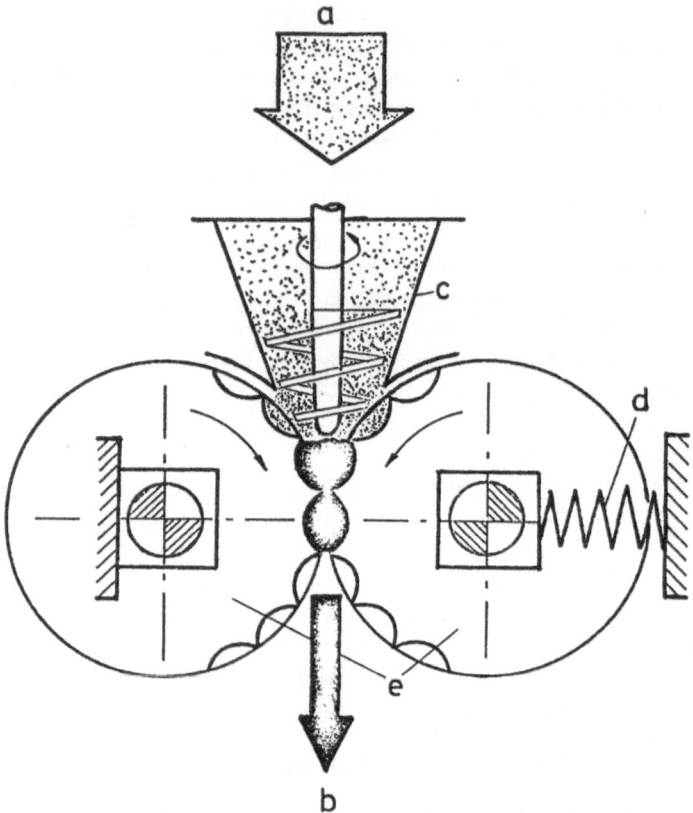

Figure 4.23 Schematic representation of the essential functioning parts of a modern roller briquetting press: a, feed; b, briquettes; c, charging device; d, spring application of pressure; e, rolls.

according to the Kelvin equation. An externally applied pressure and, in the case of thick layers, the weight of the material can also have an effect. The mechanisms of solid transport are viscous flow, evaporation and condensation, surface diffusion, volume diffusion, grain-boundary diffusion and diffusion processes favoured by displacements.

The high temperatures of the sintering process are achieved either by external heating of the charge in retorts or by oxidation reactions within the charge itself. Among the processes using external heating are, for example, the calcining of ceramic feedstocks (e.g. clinker), the sintering of cement in heated rotary kilns, the roasting of pigments and powder-metallurgical sintering in chamber or muffle furnaces, which is often carried out in an atmosphere of inert gas.

Sintering processes using internal oxidation are applied widely (e.g. the pelletizing of iron ores). Today they are carried out predominantly in suction-draught sintering plants. The sintering machine, a travelling grate, consists of a chain grate that travels around the sintering belt. The upper section between the reversing drums is coupled with a fan through suction boxes and a de-dusting device. Fine-grained moistened material mixed with the fuel is fed onto this section; subsequently it is ignited on the surface and air is sucked through. In this way the combustion zone migrates from above downwards through the charge so that the exhaust gases preheat the lower layers and, conversely, the combustion air is preheated in the completely sintered layer. To protect the grate the lowest layer on it is usually a bed of fuel-free recycled material that generally consists of a fraction screened out of the discharged sinter.

4.3.4 Coagulation

Coagulation or flocculation is the coalescence of small particles suspended in liquids or gases under the effect of Brownian molecular motion, turbulence or electrostatic forces of attraction. Coagulation in hydrosols is a classical problem of colloid physics. The colloidal particles carry surface charges of the same sign in equilibrium with the liquid, so that they repel one another and the dispersion is stabilized. The electrical forces of repulsion are greatly affected by the pH of the solution. They disappear at the isoelectric point (point of zero charge), and then only the van der Waals forces of attraction are effective. Rapid coagulation sets in, after which the liquid may become cloudy. The kinetics of this coagulation were developed by Smoluchowski in 1917.

Coagulation is a second-order reaction between reactants whose size increases. The rate constant depends on the ratio of the diameters of the coagulating particles. Smoluchowski developed the solutions for the coagulation of a monodisperse initial distribution. There is also a complete solution for continuous grain-size distributions. When the particles vary in size the rate constant is greater than that for uniform particles. The accumulation of small particles on large ones predominates over the coagulation of uniform particles. This corresponds to what is observed in practice. Coagulation in liquids is used, for example, to increase the rate of settling in sedimentation processes or, when filtering, to achieve a filter cake of better permeability. Important applications are the purification of sewage and the preparation of drinking water and boiler feed water. A large number of flocculants are available which can be used to initiate coagulation. They are either electrolytes or organic substances with branching long-chain molecules that tend to form hydrogen bonds. In special instances coagulation can also be achieved by ultrasonic or ionizing radiation.

Coagulation in gases is strongly influenced by the electrical charge on the particles. The attraction or repulsion of the particles is assisted by Coulomb forces. A second-order reaction also prevails for the electrostatic coagulation of two particles. The rate constant depends on the number of elementary charges. For particles of diameter 1 μm, for example, the rate of coagulation is increased through an opposing charge of 100 elementary charges by a factor of about 300 in comparison with uncharged particles. Coagulation in gases is used, for example, in order to separate very fine particles in a state of agglomeration in centrifugal dust removers (cyclones). Because of the low strength of the flocs, the drag forces acting on the agglomerates must be kept small in order to prevent an undesirable de-agglomeration. In cyclones this is achieved by having the gas enter with a low velocity.

4.4 COMMINUTION PROCESSES

The extraordinary diversity of the tasks to be performed by comminution has led to the development of a large number of different types of machine for crushing and grinding, which can be classified according to the following criteria.

1. The manner in which force and energy are applied.
    (a) Stressing mechanism I: stressing between solid surfaces that can be either the surfaces of the crushing implements themselves or the surfaces of neighbouring grains (stressing by pressure and friction). The stressing process can be restricted by a maximum force, a mimimum gap between the surface of the implements or by hindering the supply of energy as in the case of a ball mill. The stressing surfaces move not only perpendicularly but also tangentially to one another and their relative velocity is restricted for constructional reasons to about 10 m/s;
    (b) Stressing mechanism II: impact stressing on a solid surface which can be that of either the implement or another grain. The stressing is determined by the kinetic energy of the relative motion whose velocity lies between 20 and 200 m/s;
    (c) Stressing mechanism III: stressing by the surrounding medium which is effective, even with large shearing gradients, only in the case of weak materials such as agglomerates. De-agglomeration in very viscous media and the homogenization of emulsions and suspensions of low viscosity, e.g. milk, represent typical applications;
    (d) Stressing mechanism IV: stressing as a result of a non-mechanical supply of energy, such as by heating, electrical discharges and the like. These methods are of importance in only a few special cases;

Section 4.4 was written with the cooperation of Professor K. Schönert.

2. According to the size range of the product: machines for producing lumps larger than a few centimetres are called crushers and those for producing finer particles are called mills. We can have coarse and fine crushers or, alternatively, grit mills, grinders and pulverizers;
3. Dry and wet grinding: this distinction is justified because of the very different flow paths and constructional arrangements.

4.4.1 Crushers

Jaw and rotary crushers are suitable for breaking hard to moderately hard materials such as granite, basalt, porphyry, greenstone, limestone, ores, slags and the like. Softer materials clog the fracture gap. The reduction ratio corresponds roughly to the ratio of the maximum width of the mouth to the minimum width of the gap and approaches a value of 1 : 5 to 1 : 8. In the case of jaw crushers distinctions are made in accordance with the principles of design between toggle, single-toggle, beater and hydraulically operated crushers as well as Rotex crushers. In the case of rotary crushers, distinctions are made between cone, Symons, jacketed and rotary jaw crushers. The rotary crushers have a larger throughput than the jaw crushers but are more expensive to construct. Roller crushers can be used for hard to soft materials; soft and plastically deformable materials are crushed using knife crushers and stamp mills.

In the case of jaw crushers the feed is smashed either between a fixed and an oscillating jaw or between two oscillating jaws. The drive can be either mechanical or hydraulic. Figures 4.24 and 4.25 illustrate the principles of the Blake jaw crusher and the Dodge jaw crusher. With the former the lifter rod b oscillates around the eccentric trunnion f and moves the backing plates c which, in turn, make the swing jaw a move backwards and forwards around the point e. The forces are borne mainly by the bearing g; the bearing of the trunnion is only slightly loaded. To a first approximation the stressing surfaces move perpendicularly to each other. The gap width can be varied by shifting the bearing g. The toggle system (b, c) can exert only compressive forces and, because of this, return springs are still necessary between the swing jaw and the casing; these are not shown in the figure. Large flywheels on the shaft of the eccentric even out the intermittent demand for energy. With single-toggle crushers the motion of the eccentric shaft is transferred directly to the swing jaw which is suspended like a pendulum from the gudgeon pin d and supports the backing plate b at point f. This in turn swings on its bearing e which can be shifted not only vertically but also horizontally to adjust the crushing gap as well as the stroke at point f. A pronounced tangential movement is superimposed onto the perpendicular movement of the swing jaw towards the surface being stressed. This machine also has a return spring and a flywheel. The design of the single-toggle crusher is simpler than that of the Blake jaw crusher but, as a result, the bearing of the eccentric is more

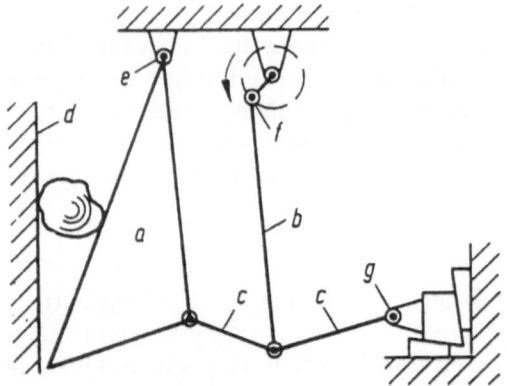

Figure 4.24 Principle of the toggle crusher: a, swing jaw; b, lifter rod; c, backing plate; d, compression plate; e, bearing; f, eccentric trunnion; g, bearing of the backing plate.

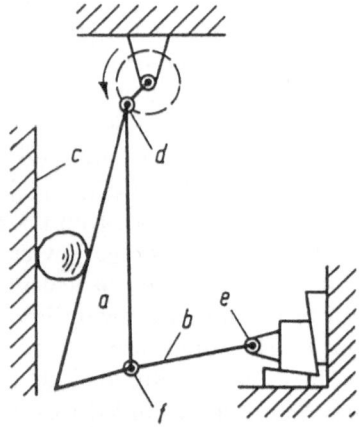

Figure 4.25 Principle of the single-toggle crusher: a, swing jaw; b, backing plate; c, compression plate; d, eccentric trunnion; e, bearing of the backing plate.

severely loaded.

The constructional principle of the cone crusher is illustrated in Fig. 4.26. The crushing cavity is bounded by an outer fixed downwardly converging conical casing and the inner crushing cone f which can rotate freely around the inclined spindle a. This spindle is supported in the bearing b and guided around a circular path by the lower bearing d. d is an eccentric hole let into the socket of the lower bearing c which is mounted at e in the housing. The width of the gap can be adjusted by shifting either the cone or the casing axially. The annular shape of the crushing cavity

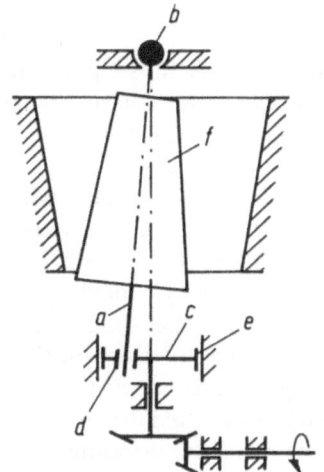

Figure 4.26 Principle of the cone crusher: a, axle of the cone; b, bearing; c, d, eccentrics; e, bearing of the eccentric; f, crushing cone.

makes it possible to obtain larger throughputs of material than is the case with the jaw crushers. The Symons crusher represents a special development which achieves larger reduction ratios of 1 : 18 up to 1 : 35.

For very soft material that is inclined to deform plastically, and especially in order not to produce too many fines, suitable machines are those with a rotor that has knife-like arms (Fig. 4.27). These arms comb through a grate and stress the lumps of material intensely at the points of contact with the arms or with the rods of the grate. With materials that behave plastically the stressing occurs mainly by shearing; if the behaviour is elastic the stressing is caused by pressure and binding. Such machines are also called toothed-roll and stamp crushers.

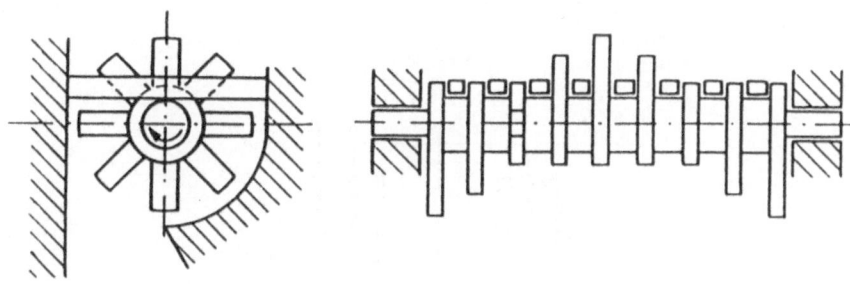

Figure 4.27 Principle of a knife crusher.

4.4.2 Roller mills

The class of roller mills includes those pulverizers in which the stressing occurs between two surfaces that roll off each other, i.e. spherical or cylindrical rollers and a smooth conical or bowl-shaped grinding face. The milling implements are enclosed in a housing. The grains form a layer of material and are stressed by pressure and shear. The contact forces are either gravitational or centrifugal, or are extraneous hydraulic or spring forces. The numerous designs of roller mills can be classified according to the following operating conditions or mechanical arrangements: the use of centrifugal or extraneous force; the shape of the rolling body (roller or ball); the orientation of the axis of the rotating rolling surface as well as the direction of pressing (radial pressing with a vertical shaft, radial pressing with a horizontal shaft, axial pressing with a vertical shaft); the drive of the rolling bodies or of the grinding face.

Roller mills are suitable for pulverizing soft to moderately hard materials down to a fineness of less than 200 μm. Because the grinding chamber is enclosed and the residence time is longer, drying can be combined with pulverizing without difficulty. Designs that can tolerate temperatures up to about 700 °C are available. The field of application of roller mills is very large and includes the dressing of ores, cement feedstock and ceramic masses and the pulverizing of coal, phosphate, gypsum, bauxite and dyestuffs. Large mills achieve throughputs of up to 150×10^3 kg/h. Roller mills are displacing ball mills for the materials listed above. Basically mills using extraneous force promise more than those using centrifugal force because with the former the pressing force and the rolling speed can be independently adjusted, whereas with the latter they are coupled together.

Of the numerous types of roller mill in existence only two will be discussed here as examples. Pendulum mills (Fig. 4.28) are centrifugal roller mills in which one or more rollers are suspended free to swing from a crosshead c. On rotation the centrifugal force presses these onto the layer of material on the grinding face. A stop prevents direct contact

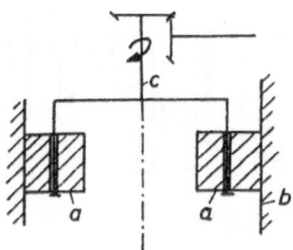

Figure 4.28 Principle of a pedulum mill: a, roller; b, grinding face; c, crosshead.

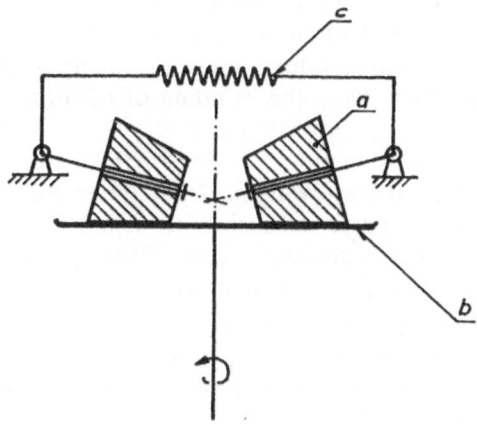

Figure 4.29 Principle of a two-roller disc mill: a, rollers; b, grinding face; c, pressing springs.

between roller and ring. Ploughshare-like vanes in front of the rollers guide the material onto the grinding face. Figure 4.29 shows the principle of an extraneous-force roller mill. The grinding face is a flat plate with an outer rim. Two or more fixed rollers are pressed against it either hydraulically or by the force of springs, and the plate is driven around.

4.4.3 Mills with loose milling implements

A very significant group of mills has freely moving milling implements that can be balls, rods, short cylindrical pieces ('Zylpebs') or the coarse grains of the feed itself (autogenous milling). These implements, which are also called tumbling bodies, are held, usually in very large numbers, within a closed container and are thrown around since either the container executes a rotating, planetary or shaking movement or a mechanical stirrer sets them in motion. The material that finds itself between the tumbling bodies is stressed either by retarding them or by their relative motion against one another under load. Mills with a rotating cylindrical or conical grinding chamber are called, according to the shape of the tumbling bodies or the container, ball, rod, trommel, cone or tube mills. The autogenous mill also belongs to this group. In it impact stressing also occurs, especially when the diameter is large (about 10 m). The other three cases mentioned above are known as planetary, vibratory or mechanically stirred mills. The last-named of these machines form a class of their own and are used almost exclusively for the milling of soft materials in suspension or to achieve de-agglomeration; they are discussed on p. 182. All other mills listed can be used for both wet and dry grinding, but mostly for the latter.

The most important field of application is the fine and extremely fine comminution of hard and moderately hard materials at high throughputs up to 200×10^3 kg/h. Furthermore, there are many instances of application in the chemical industry, from the grinding of minerals to the dispersing of fine-grained materials in highly viscous media.

Ball mills are the most important type of this group of machines. They are manufactured in all sizes ranging from laboratory mills to large-scale mills. The grinding chamber consists of a cylindrical vessel usually fitted with grinding bars (or sometimes with lifter bars) on the inside and 25%–45% filled with balls of the same or various sizes. The extent of the filling is the ratio of the bulk volume of the balls to the volume of the grinding chamber. Depending on the rotational speed and the extent of the filling three states of motion of the charge of balls can be distinguished (Fig. 4.30): cascading, cataracting and centrifuging. If the extent of the

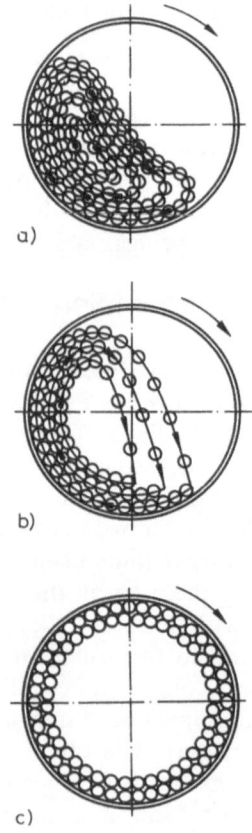

a)

b)

c)

Figure 4.30 States of motion in a ball mill: (a) cascading; (b) cataracting; (c) centrifuging.

filling is small, the charge of balls can swing backwards and forwards like a compact mass, which is a very undesirable state of affairs. The transitions between the states of motion occur readily; thus, for instance, the outer layers of balls can peel off and form a cataract while the inner layers still cascade. Centrifuging starts at the critical rotational speed n_c at which the centrifugal acceleration $\omega^2 R = 2\pi^2 n_c^2 D$ is equal to the acceleration due to gravity g ($D = 2R$ is the inner diameter of the mill). It follows that $n_c = (g/2\pi^2 D)^{1/2}$ or $n_c = 42.3/\sqrt{D}$ rev/min, where D is in metres. In practice it has proved satisfactory to set the speed at 75% of the critical value, i.e. $32/\sqrt{D}$; at this speed the balls are thrown off at an angle of about 55° below the crown of the grinding chamber. The slip between the balls and the chamber can be considerable, and hence its rotational speed gives no direct indication of their state of motion. The balls in the inner layers, which either roll off or do not come in direct contact with the wall, keep on rolling and form the 'ball train'. The motion of the balls has been investigated theoretically in various ways. Despite the considerable simplifications used, the predictions have proved to be correct to a good approximation.

The motion of the chamber of balls determines first of all the power required of the mill but it does not determine the comminution. Using energy in the best possible way to achieve comminution is a problem of adapting the mill to the material to be ground. On the one hand the impact energy of a ball must be sufficient to crush the coarsest grains, and on the other hand the resultant fines should not be briquetted again by too much compression and the number of contact points, i.e. the number of balls, should be as large as possible. Since the diameter of the chamber and the density of the balls are generally not variable, the energy must be adjusted according to the size d of the balls. However, there are no general rules for this. Open-circuit mills are equipped with classifying linings which impart an impulse to the balls towards the inlet of the mill. Larger balls thereby receive a larger impulse and collect around the inlet to the mill in a greater concentration than at the exit. Furthermore there are multichamber mills that are divided by partitions into two or three compartments. The feed passes successively through these compartments which have the appropriate charge of balls.

Planetary mills consist of a rotating carrier system on which there are one or more grinding containers that rotate around their axis with a different angular velocity. By suitable coordination of the angular velocities the tumbling bodies start moving like those in a ball mill. The centrifugal acceleration can be from 10 to 20 times that due to gravity. Smaller balls can be used, so that planetary mills are particularly suitable for very fine grinding.

An increase in the acceleration of the tumbling bodies, compared with gravitational acceleration, can also be achieved by oscillating movements.

This is exploited in vibratory mills. They consist of one or more closed circular or trough-shaped containers that are supported on springs set vibrating in a circular motion by means of a rotating eccentric weight. The containers are up to 80% filled with balls, cylindrical pebbles or rods. As a consequence of the vibrations the charge is loosened and the tumbling bodies bounce off one another and against the wall of the container, thus grinding the material between them. The whole charge rolls around in the opposite direction to the rotation, which effects a thorough mixing. Tubular vibratory mills are suitable for open-circuit grinding. They are used for the comminution of materials ranging from soft to hard, such as quartz, chamotte, limestone, argillaceous earth, slaked lime and various chemicals.

4.4.4 Impact-grinding machines

Stressing mechanism II, impact stressing on a solid surface, is realized in hammer crushers, impact crushers, beater mills and jet pulverizers. These machines have a wide range of application that is limited only by the abrasive wear which increases rapidly with the peripheral velocity. Hard materials can be pre-crushed only with peripheral velocities between 20 and 60 m/s. Impact mills for fine grinding operate with peripheral velocities up to 150 m/s or relative peripheral velocities up to 200 m/s where they have two rotors. When abrasive materials have to be ground, wear-resisting components are necessary. Plastic materials, for which compressive stressing is ineffective, can often be smashed by impact since rapid stressing allows less plastic deformation to occur. Impact grinding often leads to a better selective disintegration of multicomponent materials such as ores. Rocks are more inclined to disintegrate into cubically shaped fragments than when they are subjected to compressive stressing.

A large number of types of machine of all sizes are available from large crushers with throughputs of several hundreds of tonnes per hour down to small laboratory mills. Their constructional characteristics can be distinguished according to the following factors.

1. Number and arrangement of the rotors: one rotor, two coaxial or two parallel rotors;
2. Orientation of the rotor axle: horizontal or vertical;
3. Feed: central or peripheral;
4. Flow of the material through the device: radial, axial or peripheral;
5. Impact implements of the rotor: suspended on hinges, rigidly attached plates, noses or pins;
6. Factors limiting the fineness: in the crushing zone, grate, sieve, gap between the crushing implements or flow pattern of the stream of material; sifting outside the crushing zone but still within the mill, or outside the mill in an external classifier.

Furthermore, under factor 3 there is a variant of the rotor which shoots the feed against fixed impact plates; the rotor itself hardly stresses the grains at all.

Machines with hinged impact implements are called hammer crushers or hammer mills (Fig. 4.31). The impact implements a hang free to swing on the rotor. Their dimensions are such that they barely deflect when they encounter medium-sized grains in the feed but swing out of the way if they come against foreign bodies and particularly large and strong lumps. The feed is first thrown against the inlet chute and then against the sieve grate b until it is fine enough to escape from the crushing zone.

Some impact mill designs are shown in Fig. 4.32. The variants shown in Figs 4.32(a)–4.32(d) can be termed hammer-bar mills with cylindrical, V-shaped or cone-shaped crushing chambers. The feed flows in at the centre and must pass in spirals through the gap between the rotor fitted with bars and the profiled stator. In Fig. 4.32(a) the material is classified by a perforated basket, in Fig. 4.32(b) by the spiral flow between the back of the rotor and the housing, and in Fig. 4.32(d) by the spiral flow in the cylindrical space at the rear. The type shown in Fig. 4.32(f) has pins as the impact implements; there are also pin mills with counter-rotating pins. The partial diagram in Fig. 4.32(g) shows a centrifuging mill; the grains are stressed on the outer rim of the impact plates.

The important factors in the operation of impact mills are the peripheral velocity, the clearance between rotor and stator, the shape, size and number of impact implements, the feed concentration, the temperature and moisture content of the carrier gas and the way in which the classifying

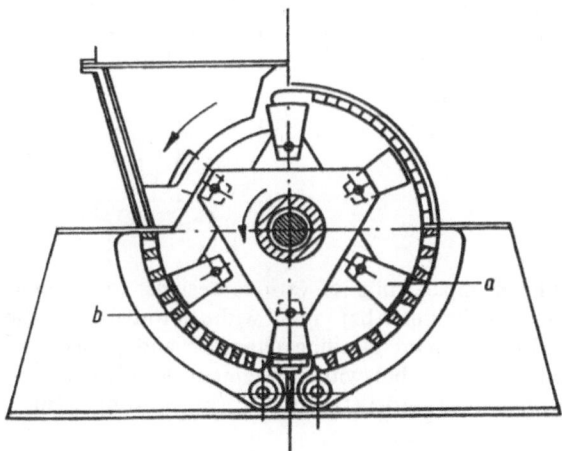

Figure 4.31 Principle of a hammer crusher with a sieve grate: a, impact implement; b, sieve grate.

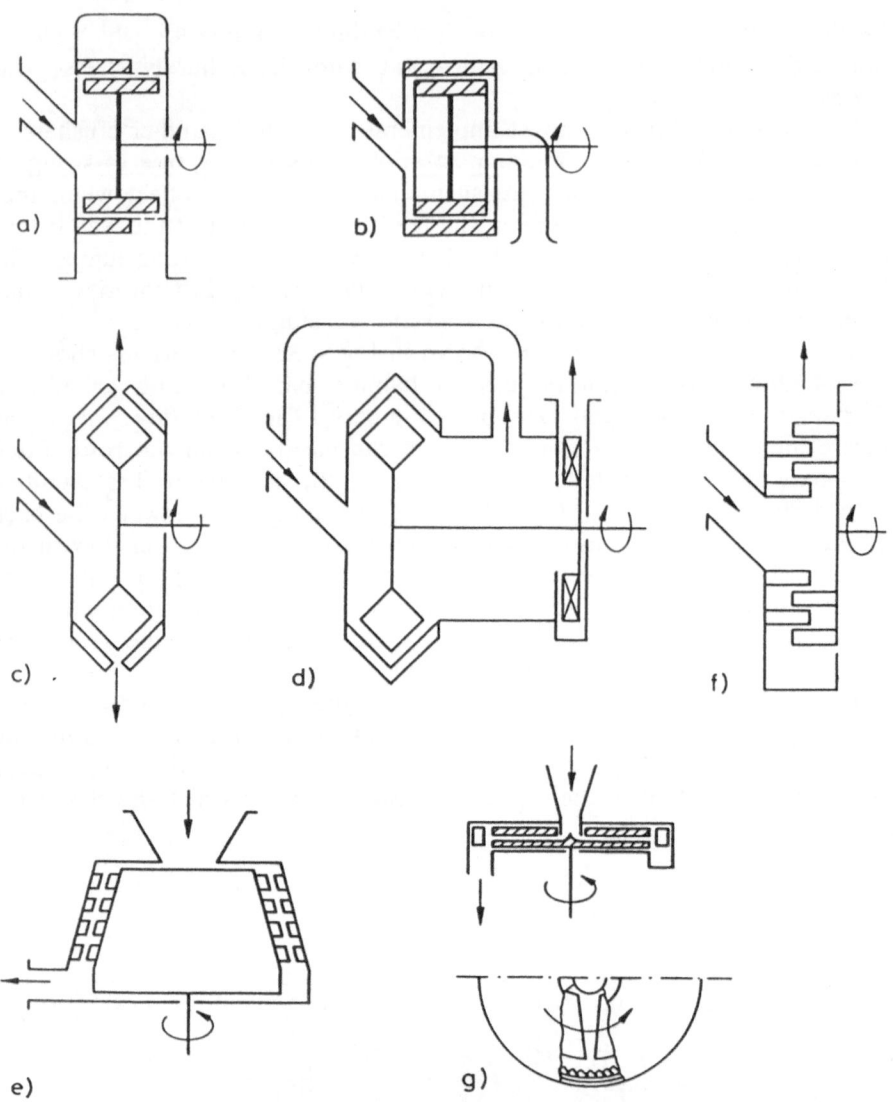

Figure 4.32 Principle of impact mills of various designs: (a) hammer-bar mill with sieve classification; (b) hammer-bar mill with aerodynamic classification; (c) hammer-bar mill with V-shaped grinding chamber and peripheral escape of the ground material; (d) as for (c) but with subsequent aerodynamic classification; (e) hammer-bar mill with conical rotor; (f) pin mill; (g) centrifuging mill.

devices influence the residence time and the fineness achieved. The types of machines are distinguished in this respect, as are the possible variations. The best design for a given comminuting task therefore has to be chosen.

A jet mill consists of a grinding chamber into which jets of the gaseous propellant medium are blown at high velocities, usually sonic or supersonic, and the feed is introduced by means of an injector. The grains are accelerated into the jets and collide with one another or with the wall. Which of the two effects prevails depends on the state of the flow which can be regulated by varying the inlet pressure or by changing the direction of the jet. Figure 4.33 shows two common kinds of jet mill. There are also other versions with two opposing solids-loaded jets. The escaping stream of propellant is sifted. The energy required by jet mills is very great and they are used for very fine grinding or for materials that must not be contaminated.

4.4.5 Cutter mills

Scrap and foil from tough plastics, rubber, leather and foodstuffs cannot be ground in the mills discussed so far. For this purpose machines must be chosen that apply stress by cutting. The same is true for the disintegration of plant material in sheaves of fibre and the production of wood chips and cellulose pulp. Special disintegrators have been introduced for such tasks: cutter mills, chippers, knife cutters, toothed disc mills and, particularly for reduction to fibres, the pulping machine.

Cutter mills were developed in order to reduce plastics to granules

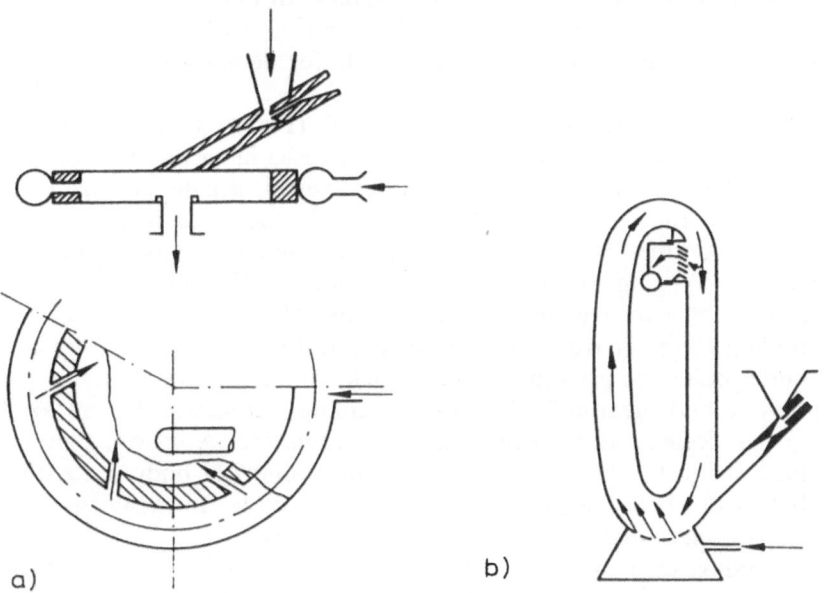

a)   b)

Figure 4.33 Principle of jet mills of two designs: (a) spiral jet mill; (b) oval-tube jet mill.

(grains between 0.05 and 0.5 cm). In extrusion granulators the bands of plastic coming out of the nozzles are cut up into strips after cooling and introduced into a cutting device consisting of a stator–rotor cutter. A uniformly cubical granulate is made from the band. Cutter mills are constructed like jet mills. The rotor and the stator are furnished with knives that pass one another with a narrow gap. The peripheral velocities lie in the range 5–20 m/s. The screening basket determines the ultimate fineness of the material.

4.4.6 Wet milling

A great many materials must be milled in liquid or paste suspensions. The viscosity of the suspension medium can vary over a range of several orders of magnitude: from 10^{-2} P for water to 10^4 P for highly viscous oils. As well as depending on the medium, the viscosity of suspensions also depends on the concentration and fineness of the solid; comminution can therefore change the viscosity. Not only solid materials like ores but also loose agglomerates of dyestuffs are broken up. In the field of wet milling very different types of machines are therefore encountered, including homogenizers with high-speed rotors, ball mills, stirred ball mills, roll stands, colloid mills and kneaders.

Machines with fast-running profiled or toothed rotor–stator systems and equipment in which suspensions are forced through narrow gaps under great pressure are designated colloid mills, high-speed stirrers and homogenizers. These machines are suitable for de-agglomeration in liquids of low viscosity.

In recent decades the stirred ball mill (Fig. 4.34) has become well established for many milling and dispersing tasks in suspensions of medium viscosity. This mill consists of a cylindrical vessel, usually arranged with its axis vertical, that is completely filled (merely indicated in Fig. 4.34) with sand grains, steel balls, steatite balls or similar grinding media with dimensions ranging from a few tenths of a millimetre up to several millimetres. A stirrer is installed in the axis. The suspension flows through the milling chamber from the bottom to the top (*sic*). A sieve or a slit at the outlet holds the grinding medium back. Comminution occurs as a result of pressure and shearing between the grinding media. Stirrers are available in various designs: discs, rings, flat steel arms etc. A special type of this machine is the blocked-ball mill. Here the cylindrical body is rotated and the ball is blocked and turned back at a fixed plate. In many cases stirred ball mills replace ordinary ball mills and roll stands.

Roll stands with two or more smoothly ground rollers are used for the processing of moderately viscous suspensions, e.g. for the grinding of pigments or for the extremely fine comminution of chocolate pastes. The rollers run at different speeds, each one faster than the preceding one.

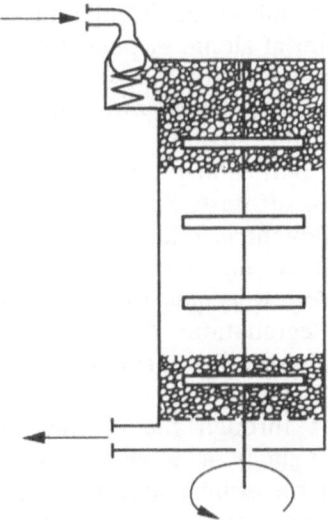

Figure 4.34 Principle of a stirred ball mill.

This effects the transport of the layer of suspension. In many machines a plate, generally known as a bar, replaces one of the rollers. The gap is adjusted by varying the contact forces, which are generally applied hydraulically. The smallest gap has a width of 20 μm. Coarse grains are stressed, depending on the setting of the gap, by pressure between the rollers; fine grains are stressed similarly if there is a large enough concentration to give rise to grain bridges. In other cases only the shear stress transferred by the medium is effective. Generally several passes through the rollers are necessary to achieve the fineness desired.

4.5 CONVEYING, STORING AND FEEDING OF BULK SOLIDS

4.5.1 Conveying

Conveying procedures can be classified according to the medium to be conveyed and the mode of operation of the conveying equipment (continuous or intermittent). Examples of continuously operating conveying plants for bulk solids are continuous mechanical conveyors such as belt, screw and vibrating conveyors, bucket elevators and Redler conveyors. In addition to these examples of the principle of continuous transport there is pneumatic and hydraulic conveying. Only pneumatic conveying will be discussed in any detail here.

Section 4.5 was written with the cooperation of Dr H. Reichert.

Pneumatic conveying of bulk solids through ducts takes place in a stream of air that moves the material along. A pneumatic conveying plant consists essentially of the feeding device for the solids, a conveying pipe, a solids separator and a blower. In contrast with other methods of conveying, pneumatic conveying has the advantage that the conveying line is completely enclosed. This is particularly important in the case of very fine-grained materials that give off dust. Some of its disadvantages are that, as a rule, the power requirement is relatively high (1–4 kWh per tonne of conveyed material) and that, under certain circumstances, the conveying equipment undergoes severe wear and the conveyed material may suffer a certain amount of size degradation. A distinction is made between air-borne, displacement and fluidization conveying.

In the most common method, i.e. air-borne conveying which permits completely free transport through the channel, the conveying medium carries the particles with it singly or in 'clouds'. In vertical ducts and at low volume concentrations of the solid material the particles distribute themselves almost uniformly over the cross-section; strands appear only at heavier loadings. In horizontal or inclined ducts only very fine dusts at low concentrations are homogeneously distributed. As the particles become larger or the loading increases the concentration in the cross-section of the duct is higher at the bottom that at the top so that a stratified flow finally appears (transitional state between air-borne and displacement conveying): in the lower part of the duct the strands of dust move with a high solids content and low velocity and above these there is a solid–gas mixture with a low dust loading and a higher velocity. If the solids loading is increased still further, part of the dust is deposited in dune-like formations at the bottom or in a uniform layer. Above this a strand of dust moves relatively slowly and, in the upper part of the duct, a quasi-homogeneous dust–gas mixture flows with high velocity. The limit is reached as the duct continues to fill up and become clogged, particularly at places where the dust is inclined to deposit (horizontal conduits directly downstream from feed points, branches or bends).

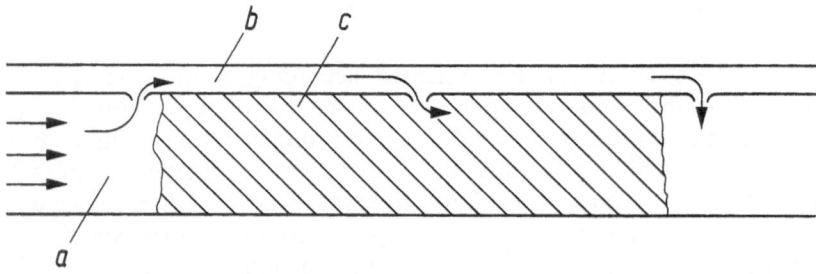

Figure 4.35 Schematic arrangement of a section of a pneumatic conveyor for displacement conveying: a, conveying duct; b, bypass; c, plug.

In displacement or dense-phase conveying, densely packed material is pushed through the conduit in the form of plugs. Blockage is avoided by breaking up a long plug into many short ones. This is achieved by a supplementary or bypass conduit which, at intervals of about ten times its diameter, has holes linking it with the conveying duct (Fig. 4.35). As a result of the pressure differential between the bypass and the duct conveying the plug of material, air flows from the bypass into the middle of the plug and breaks it up into two short plugs. Displacement conveying exhibits several advantages compared with air-borne conveying; for example, at high dust concentrations it is easier to separate the solid material, the duct suffers less erosion and the material is more carefully handled.

In fluidized conveying the flowability of fine-grained materials is improved by fluidizing them with air. This kind of conveying requires there to be a slope in the duct but, *inter alia*, it makes it possible to change the direction and inclination of the flow in any way. In fluidized conveying the energy required is usually less than that for continuously operating mechanical conveyors.

A large amount of empirical data is available for the design of pneumatic conveyors.

4.5.2 Storing and feeding

Silos and bunkers – the two terms are regarded as equivalent – are used as containers for the storage of bulk solids. They are built both with and without discharge hoppers. The cross-section is usually round, square or rectangular. In designing a bunker of suitable shape one starts with the flow properties of the bulk solid. When it flows down the bunker either mass flow or core flow (piping) can prevail. In any event one must attempt to achieve mass flow so that the whole charge moves; when core flow occurs some of the content remains stationary, so that the volume of the bunker is only partly utilized and the bulk solid can cake together in the dead spaces and be spoilt. Whether mass flow occurs is determined by the conditions in the discharge. Discharge hoppers can be successfully designed using Jenike's method (see section 3.4.4). This often leads to very slender silos.

The capacity of a bunker can often best be raised by increasing a horizontal dimension. Then several outlets, which can be opened either successively or simultaneously, are necessary. For large storage capacities silo plants consisting of groups of single silos are frequently used. Such a plant also allows, under mass flow, a bulk solid to be homogenized in that specified amounts of material can be withdrawn in a certain sequence.

Flat-bottomed bunkers, i.e. those with no conical outlet or with an outlet that is too flat to drain, have the advantage that, for a given volume, less height and therefore less internal space is required. However, since the bulk solid will not automatically flow out completely, removal of the charge must be facilitated by mechanical or pneumatic means. Such bunkers are used for the storage of fine powders that must be fluidized in order to extract them, and where there are constructional constraints that do not permit tall vessels or when the cost of building discharge hoppers is to be avoided; here a careful cost comparison which takes into account the greater advantages of spontaneous mass flow is, of course, recommended.

In the case of bunkers with discharge hoppers the shape of the hopper is of great importance with respect to the operation of the plant since the ability of the solid to flow in this part of the bunker is largely determined by the form of the hopper. The relevant controlling factors are the size of the outlet hole and the slope of the hopper. To avoid piping, round, square or rectangular hoppers with a slotted outlet are also used, with the length of the slot being the same as the diameter of the bunker.

The additional contrivances that are necessary for the bunkering of bulk solids are discharging devices and discharging aids. Discharging devices are those pieces of equipment that feed the material coming out of the bunker at a controlled rate to the next step in the process. Discharging aids are devices and machines that serve to maintain the flow of material in and out of the bunker and to improve the profile of this flow. This occurs frequently in conjunction with the discharging device which can itself also act as a discharging aid. In a broad sense any special constructional shaping of the hopper and procedures that alter the properties of the bulk solid can be regarded as discharging aids.

In practice discharging aids are used with flat bunkers for the reasons indicated above. Often bunkers are also constructed without any knowledge of the properties of the material to be stored in them. It is then found during operation that blockages occur in the discharge. A retrofit of discharging aids is then supposed to ensure that the bunker will still function in the desired way. If different bulk solids with varying properties are to be stored in a bunker, discharging aids are installed in order to ensure that the materials with the poorest flow properties will flow out of it.

Among the discharging aids that change the properties of material are dispersants that consist of very fine particles of sizes from about 0.1 to 0.01 μm and which, when added to the bulk solid, improve its flow properties because the adhesive forces between its particles are reduced (see p. 118). Another device that assists discharge is the injection of air through a bunker with a porous bottom. In this way the material is loosened up to a certain height. Other discharging aids include shakers that are flanged onto the outside of the wall of the bunker or inserts that are positively moved (stirrers, vibrating grates), suspended in position (e.g. a cone or beam in the hopper) or set in motion by the flowing material (flexible walls, loosely hanging chains).

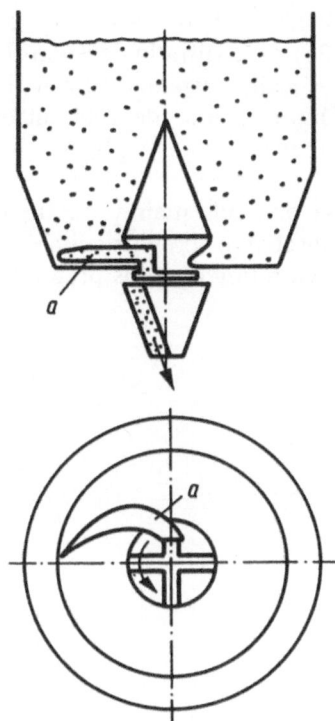

Figure 4.36 Silo with a flat bottom and a rotating discharging arm a.

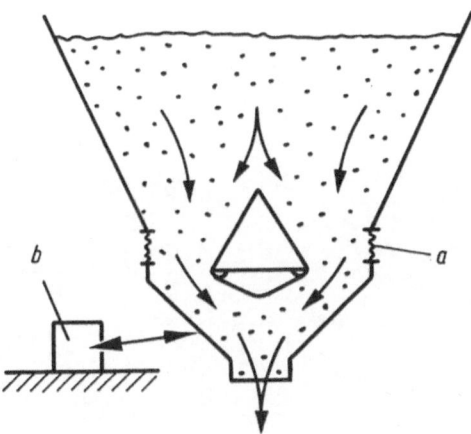

Figure 4.37 Vibrating discharging system: a, elastic suspension; b, motor driving an unbalanced flywheel.

Dosing devices, which also act as discharging aids, include the rotating discharging arm (Fig. 4.36) or a vibrating system that sets the lower part of the hopper, which is elastically suspended from the rest of the hopper, in vibration (Fig. 4.37). Other dosing devices include ordinary conveyors, plate conveyors, vane feeders, rotary plates, trolleys, discharge screws, shaking tables and vibrating conveyors. Which dosing device can be used for which particular case depends mainly on the particle size of the bulk solid. For instance, the discharge screw is suitable mainly for fine and very fine materials, while the vibrating chute is preferred for coarse particles.

References

BOOKS

1. Allen, T. (1968) *Particle Size Measurement*, Chapman and Hall, London.
2. Batel, W. (1971) *Einführung in die Korngrößenmeßtechnik*, 3rd edn, Springer, Berlin.
3. Brauer, H. (1971) *Grundlagen der Einphasen- und Mehrphasenströmungen*, Sauerländer, Aarau-Frankfurt-am-Main.
4. Carman, P. C. (1956) *Flow of Gases through Porous Media*, Butterworth, London.
5. Darcy, H. P. G. (1856) *Les Fontaines Publiques de la Ville de Dyon*, Dalmont, Paris.
6. Grassmann, P. (1970) *Physikalische Grundlagen der Verfahrenstechnik*, 2nd edn, Sauerländer, Aarau-Frankfurt-am-Main.
7. Gupte, A. R. (1970) Experimentelle Untersuchung der Einflüsse von Porosität und Korngrößenverteilung im Widerstandsgesetz der Porenströmung, Dissertation, Universität Karlsruhe.
8. Herdan, G. (1960) *Small Particle Statistics*, 2nd edn, Butterworth, London.
9. Hoerner, S. F. (1958) *Fluid-Dynamic Drag*, Midland Park, NJ.
10. Kaskas, A. A. (1970) Schwarmgeschwindigkeiten in Mehrkornsuspensionen am Beispiel der Sedimentation, Dissertation, Technische Universität Berlin.
11. Kaye, B. H. and Boardman, R. P. (1962) Cluster formation in dilute suspensions. *3rd Congress of the European Federation of Chemical Engineers*, A 1721, Institution of Chemical Engineers, London.
12. Kendall, M. G. and Moran, P. A. P. (1963) *Geometrical Probability*, Griffin, London.
13. Koglin, B. (1971) Untersuchungen zur Sedimentationsgeschwindigkeit in niedrig konzentrierten Suspensionen, Dissertation, Universität Karlsruhe.
14. Koglin, B. (1972) Settling rate of individual particles in suspension. In M. J. Groves and J. L. Wyatt-Sargent (eds), *Particle Size Analysis 1970*, p. 223, Society for Analytical Chemistry, London.
15. Koglin, B. (1973) Dynamic equilibrium of settling velocity distribution in dilute suspensions of spherical and irregularly shaped particles. *Proceedings of the International Conference in Particle Technology, Chicago 1973*.
16. Koglin, B. (1973) Methods for the determination of the degree of agglomeration in suspensions. *Proceedings of the International Conference in Particle Technology, Chicago 1973*.
17. Krupp, H. (1971) *Static Electrification*, Institute of Physics, London.

18. Parfitt, G. D. (1969) *Dispersion of Powders in Liquids*, Elsevier, Amsterdam.
19. Raasch, J. (1962) Die Bewegung und Beanspruchung kugelförmiger und zylindrischer Teilchen in zähen Scherströmungen. In D. Behrens (ed.), *Einige Neuentwicklungen von Fein-und Feinstprallzerkleinerungsmaschinen*, p. 138, Verlag Chemie, Weinheim/Bergstraße, VDI-Verlag, Düsseldorf.
20. Rumpf, H. and Schönert, K. (1972) *Die Brucherscheinungen in Kugeln bei elastischen sowie plastischen Verformungen durch Druckbeanspruchung*, Dechema Monograph 69, p. 51, Verlag Chemie, Weinheim/Bergstraße.
21. Sawatzki, O. (1961) Über den Einfluß der Rotation und der Wandstöße auf die Flugbahnen kugliger Teilchen im Luftstrom, Dissertation, Universität Karlsruhe.
22. Schlichting, H. (1965) *Grenzschicht-Theorie*, 5th edn, Braun, Karlsruhe.
23. Schubert, H. (1968, 1972) *Aufbereitung fester mineralischer Rohstoffe*, Vols 1 and 2, 2nd edn, 1968; Vol. 3, 1972, VEB Deutsch. Verl. Grundstoffindustrie, Leipzig.
24. Schubert, H. (1972) Untersuchungen zur Ermittlung von Kapillardruck und Zugfestigkeit von feuchten Haufwerken aus körnigen Stoffen, Dissertation, Universität Karlsruhe.
25. Schwedes, J. (1968) *Fließverhalten von Schüttgütern in Bunkern*, Verlag Chemie, Weinheim/Bergstraße.
26. Schwenk, W. (1972) *Oberflächenveränderungen von Feststoffen nach Zerkleinerung im Hochvakuum*, Dechema Monograph 69, p. 121, Verlag Chemie, Weinheim/Bergstraße.
27. Tchen, C. M. (1947) Mean value and correlation problems connected with the motion of small particles suspended in a turbulent fluid, Dissertation, University of Delft.
28. Ullrich, H. (1967) *Mechanische Verfahrenstechnik*, Springer, Berlin.
29. Winnacker, K. and Küchler, L. (eds) (1974) Chemische Technologie, Carl Hanser, Munich.

JOURNAL PUBLICATIONS

30. Alex, W. (1972) *Aufbereit. Tech.*, **13**, 105, 168, 639, 723.
31. Barfod, N. (1972) *Powder Technol.*, **6**, 39.
32. Batchelor, G. K. (1972) *J. Fluid Mech.*, **52**, 245.
33. Brauer, H. and Kriegel, E. (1965) *Chem.-Ing.-Tech.*, **37**, 265.
34. Brauer, H. and Kriegel, E. (1966) *Chem.-Ing.-Tech.*, **38**, 321.
35. Brenner, H. (1963, 1964) *Chem. Eng. Sci.*, **18**, 1; **19**, 599.
36. Brenner, H. and Happel, J. (1958) *J. Fluid Mech.*, **4**, 195.
37. Brezina, J. (1969) *J. Sediment. Petrol.*, **39**, 1627.
38. Burgers, J. M. (1941, 1942) *Proc. Ned. Akad. Wet.*, **44**, 1045, 1177; **45**, 9, 126.
39. Crowley, J. M. (1971) *J. Fluid Mech.*, **45**, 151.
40. Cunningham, E. (1910) *Proc. R. Soc. Lond. Ser. A.*, **83**, 357.
41. Davies, C. N. (1945) *Proc. Phys. Soc.*, **57**, 259.
42. Dennis, S. C. R. and Walker, J. D. A. (1971) *J. Fluid Mech.*, **48**, 771.
43. Dryden, H. L., Schubauer, G. B., Mock, W. C. and Skramstad, H. K. (1937) *NACA Rep. 581*.

44. Ergun, S. (1952) *Chem. Eng. Prog.*, **48**, 89.
45. Famularo, J. and Happel, J. (1965) *AIChE J.*, **11**, 981.
46. Goldman, A. J., Cox, R. G. and Brenner, H. (1966) *Chem. Eng. Sci.*, **21**, 1151.
47. Grassmann, P. (1961) *Chem.-Ing.-Tech.*, **33**, 348.
48. Hamaker, H. C. (1937) *Physica*, **4**, 1058.
49. Hamielec, A. E., Hoffman, T. W. and Ross, L. L. (1967) *AIChE J.* **13**, 212.
50. Hasimoto, H. (1959) *J. Fluid Mech.* **5**, 317.
51. Hocking, L. M. (1964) *J. Fluid Mech.* **20**, 129.
52. Ihme, F., Schmidt-Traub, H. and Brauer, H. (1972) *Chem.-Ing.-Tech.*, **44**, 306.
53. Jenike, A. W. (1964) *Bull. Univ. Utah*, **53** (26), 1.
54. Johne, R. (1966) *Chem.-Ing.-Tech.*, **38**, 428.
55. Koglin, B. (1971) *Chem.-Ing.-Tech.*, **43**, 761.
56. Koglin, B. (1972) *Chem.-Ing.-Tech.*, **44**, 515.
57. Kozeny, J. (1927) *Sitzungsber. Akad. Wiss. Wien, Math.-Naturwiss. Kl., Abt. 2A,`136*, 271.
58. Krekel, J. (1966) *Chem.-Ing.-Tech.*, **38**, 229.
59. Krupp, H. (1967) *Adv. Colloid Interface Sci.*, **1**, 111.
60. Kürten, H., Raasch, J. and Rumpf, H. (1966) *Chem.-Ing.-Tech.*, **38**, 941.
61. Kynch, G. J. (1959) *J. Fluid Mech.*, **5**, 193.
62. Ladenburg, R. (1907) *Ann. Phys.*, **23**, 447.
63. Leschonski, K. and Johne, R. (1966) *Informationsdienst Arbeitsgem. Pharmaz. Verfahrenstech.*, **12**, 1.
64. Lewis, W., Gilliland, E. R. and Bauer, W. C. (1949) *Ind. Eng. Chem.*, **41**, 1104.
65. Lifshitz, E. M. (1956) *Sov. Phys.—JETP*, **2**, 73.
66. Löffler, F. and Muhr, W. (1972) *Chem.-Ing.-Tech.*, **44**, 510.
67. Molerus, O. (1967) *Chem.-Ing.-Tech.*, **39**, 341.
68. Molerus, O., Pahl, M. H. and Rumpf, H. (1971) *Chem.-Ing.-Tech.*, **43**, 376.
69. Mühle, J. (1972) *Chem.-Ing.-Tech.*, **44**, 889.
70. Newitt, D. M. and Conway-Jones, J. M. (1958) *Trans. Inst. Chem. Eng.*, **36**, 422.
71. Ohe, W. von der (1967) *Chem.-Ing.-Tech.*, **39**, 357.
72. Pahl, M. H., Schädel, G. and Rumpf, H. (1973) *Aufbereit.-Tech.*, **14**, 5, 10, 11.
73. Paris, P. C. and Sih, G. C. M. (1964) *Am. Soc. Test. Mater. Spec. Publ., 381*, 30.
74. Pawlowski, J. (1964) *Chem.-Ing.-Tech.*, **36**, 1089.
75. Pietsch, W. and Rumpf, H. (1967) *Chem.-Ing.-Tech.*, **39**, 885.
76. Pyun, C. W. and Fixman, M. (1964) *J. Chem. Phys.*, **41**, 937.
77. Reichert, H. (1973) *Chem.-Ing.-Tech.*, **45**, 391.
78. Rumpf, H. (1961) *Materialprüfung*, **3**, 253.
79. Rumpf, H. (1965) *Chem.-Ing.-Tech.*, **37**, 187.
80. Rumpf, H. (1973) *Aufbereit.-Tech.*, **14**, 59.
81. Rumpf, H., Alex, W., Johne, R. and Leschonski, K. (1967) *K. Ber. Bunsen Ges. Phys. Chem.*, **71**, 253.
82. Rumpf, H. and Ebert, K. F. (1964) *Chem.-Ing.-Tech.*, **36**, 523.

83. Rumpf, H. and Gupte, A. R. (1971) *Chem.-Ing.-Tech.*, **43**, 367.
84. Schönert, K. and Steier, K. (1971) *Chem.-Ing.-Tech.*, **43**, 773.
85. Schubert, H. (1973) *Chem.-Ing.-Tech.*, **45**, 396.
86. Schubert, H. (1974) *Chem.-Ing.-Tech.*, **46**, 333.
87. Smekal, A. G. (1955) *Anz. Österr. Akad. Wiss. Math.-Naturwiss. KI*, 72.
88. Smoluchowski, M. S. (1911) *Bull. Int. Acad. Sci. Cracovie*, **1A**, 28.
89. Stange, K. (1954) *Chem.-Ing.-Tech.*, **26**, 331.
90. Stimson, M. and Jeffery, G. B. (1926) *Proc. R. Soc. Lond. Ser. A*, **111**, 110.
91. Stokes, G. G. (1850) *Trans. Camb. Phil. Soc.*, **9**, 8.
92. Taneda, S. (1956) *J. Phys. Soc. Jpn*, **11**, 1104.
93. Torobin, L. B. and Gauvin, W. H. (1959) *Can. J. Chem. Eng.*, **37**, 224.
94. Torobin, L. B. and Gauvin, W. H. (1961) *AIChE J.*, **7**, 615.
95. Trawinski, H. F. (1964) *Keram. Z.*, **16**, 20.
96. Tromp, K. F. (1937) *Glückauf*, **73**, 125, 151.
97. Visher, W. M. and Bolsterli, M. (1972) *Nature* 239 (5374), 504.
98. Wadell, H. (1933) *J. Geol.*, **41**, 310.
99. Wegener, P. P. and Ashkenas, H. (1961) *J. Fluid Mech.*, **10**, 550.
100. Wen, C. Y. and Yu, Y. H. (1966) *Chem. Eng. Prog. Symp. Ser.* 62.
101. Werther, J. (1973) *Chem.-Ing.-Tech.*, **45**, 375.
102. Wicksell, S. D. (1925) *Biometrika*, **17**, 84.
103. Wolf, K. and Rumpf, H. (1941) *VDI Z.*, Beinheft Verfahrenstechn, **2**, 29.

Index